Images of Modern America

LOST COAL COUNTRY OF NORTHEASTERN PENNSYLVANIA

Front Cover: A coal cart is filled with anthracite in Weatherly, Pennsylvania. In the background, a Hollywood movie set from *The Molly Maguires* welcomes visitors to Eckley Miners' Village. (Author's collection.)

Upper Back Cover: A railcar from the famed Erie Lackawanna Railroad is on display at Steamtown National Historic Site in Scranton. (Author's collection.)

Lower Back Cover (from left to right): On the left is an interior view of the St. Nicholas Breaker, near Mahanoy City. In the center, the Huber Breaker in Ashley was demolished in 2014. On the right, the coal mines of Northeastern Pennsylvania brought prosperity and tragedy to the region. (All, author's collection.)

Images of Modern America

LOST COAL COUNTRY OF NORTHEASTERN PENNSYLVANIA

Lorena Beniquez

ARCADIA PUBLISHING

ISBN 978-1-4671-2641-0

Published by Arcadia Publishing
Charleston, South Carolina

Printed in the United States of America

Library of Congress Control Number: 2017936455

For all general information, please contact Arcadia Publishing:
Telephone 843-853-2070
Fax 843-853-0044
E-mail sales@arcadiapublishing.com
For customer service and orders:
Toll-Free 1-888-313-2665

Visit us on the Internet at www.arcadiapublishing.com

This book is dedicated to my great-grandfather Stefano Pantano, who was a coal miner in Northeastern Pennsylvania.

Contents

Acknowledgments

This book would not have been possible without the interviewees who gave of their time and knowledge selflessly. Thanks are owed to the following people: Tom Supey Jr., retired coal miner and guide from the Lackawanna Coal Mine; Ray Clarke and William C. Best from the Huber Breaker Preservation Society; Richard Morgan from Reading Anthracite Corporation; Eckley Miners' Village historic site administrator Dr. Bode Morin; coal miner descendant Leonard Kozick; Paul Coombe from the Mahanoy Historical Society; and Eastern Pennsylvania Coalition for Abandoned Mine Reclamation executive director Robert Hughes.

A thank-you must also be extended to Kelly Mifsud and J.R. Frisbee for stepping in to help me fill crucial holes in the book. Without you two, this would have been a pamphlet, not a book.

Finally, a thank-you goes to my family for their support. To my husband, Chris Murnane, you have had to put up with the whims of an artist yet remained supportive throughout the process of compiling a book. To my mother, Sandra Shumway, thank you for pointing me in the right direction, and to my father, Jose Beniquez, I promise you have not heard the last about coal from me.

Unless otherwise noted, all images are from the author's collection.

INTRODUCTION

"It's a miracle that people came out as good as they did in this area considering what they had to put up with. The agony, disappointment, aggravation, and heartache they had to live through," says Tom Supey, who is a third-generation miner from Northeastern Pennsylvania.

Life in the anthracite industry was miserable for most. Each day, they awoke in homes that were akin to living conditions in Third World countries. Some walked for miles to their jobs, where working conditions were inhumane. For miners, if they managed to escape injury or death on the job, many later contracted black lung disease, which would eventually slowly suffocate them.

Life was no easier for the families of those in the anthracite industry. Wives struggled to raise families once they became widows. Boys were sent to work in breakers as young as eight years old. This meant no education and no hope of escaping the anthracite fields.

Many of those who worked in the mining industry were immigrants. They had left their families for a better life in America only to find their new lives were worse than what they had left behind.

It is no wonder so many families in Northeastern Pennsylvania were too ashamed and traumatized to share their family history with younger generations. Social ills spilled out from the mines affecting generations to come. "They were working in a business that they couldn't get out of and they drank to forget their troubles. They would come home, beat their wives, their kids and then go to bed," says Supey.

Families were ripped apart in other devastating ways. For instance, the 1896 Twin Shaft disaster in Pittston scarred so many. According to Explorepahistory.com, 58 men and boys were killed when a mine collapsed, taking down 200 acres of land. The tragedy left 101 children fatherless and 31 wives widowed.

Outside of the mines, additional dangers faced those in the anthracite industry. The Molly Maguires was a secret Irish society in Northeastern Pennsylvania whose members fought against coal company practices using violence. Paul Coombe with the Mahanoy Historical Society tells of a brutal murder that occurred in Wiggins in retaliation to the Maguires' tyranny. "There was a Molly Maguire incident that happened in the 1870s. At the O'Donnell house, in the middle of the night, masked men came into the house and shot a pregnant woman and a man. It was thought to be an attribution killing for what the Molly Maguires had done. Nobody ever went on trial for the crime," says Coombe.

Perhaps one of the most disturbing revelations about Northeastern Pennsylvania's past is the prevalence of corruption. "When I saw *The Sopranos* series come out on television, I was not shocked by it. I lived through this. This was prevalent in the Wyoming Valley and Scranton area; the Mafia, and being under the gun and a guy coming to your business every week and says, 'Hey, do you have my envelope ready?'" says Supey.

Those were dark times in the region's history, and as a result, nobody wanted to talk about the past. Folks just wanted to move on and look to a hopefully better future. That is understandable, but unfortunately, many families lost the benefit of gaining an oral history of their ancestors and of their nation.

Time has started to heal wounds of the past, and many are now exploring the region's history. Thankfully, there are many who are making sure we won't forget our past. The interviewees in this book are voices we need to hear. Since there are no miners left from the heyday of coal, someone needs to remind us from whence we came.

There are still some physical reminders of anthracite's past that dot the coalfields of Northeastern Pennsylvania. Unfortunately, they are disappearing, and not all of them can be saved. "You can't preserve everything. Industrial heritage poses a challenge that others don't because of the scale of them and some of the toxins that go with them," says Dr. Bode Morin, who is the historic site administrator for the Pennsylvania Anthracite Heritage Museum and Scranton Iron Furnaces. In addition, he is the interim site administrator for Eckley Miners' Village.

The Huber Breaker in Ashley is one such industrial site that was not able to be saved. Locals fiercely attempted to preserve the breaker by turning it into a museum. However, it would have taken millions of dollars to do so.

The St. Nicholas Breaker near Mahanoy City is now the last breaker in the state. Like with the Huber, efforts were made to save the St. Nicholas but were unsuccessful. It is currently undergoing demolition with no hope of being saved.

One way of preserving these anthracite relics for future generations is through photography. That is what prompted me to undertake this book. In addition to being a history refresher, this is a tour guide of sorts pointing the curious toward our past.

When visiting the historic sites, there is a sense of wonder and a bit of survivor's remorse. For instance, the Tunkhannock Viaduct in Nicholson is a testament to our region's ingenuity, and you can't help but marvel at its engineering and size. However, when you visit the site of the Knox Mine disaster, the sadness for those miners who lost their lives under the Susquehanna River can be overwhelming.

There is one positive that came from the gloom of Northeastern Pennsylvania's past. Anthracite gave the region and the nation strength. If coal was the fire of the Industrial Revolution, then miners were its flame keepers. Because of those miners, the nation had heat, won wars, and remained a world power.

The spirit of the men who built America is history worth repeating. William C. Best, of Huber Breaker Preservation Society, succinctly says, "I take every opportunity to go back and see what it was like for them. I don't ever minimize that. I believe there is a spirit of a miner that needs to be told and carried on."

One

Pennsylvania's Last Coal Breaker

Hundreds of coal breakers populated the terrain of Northeastern Pennsylvania beginning in the 1800s. Today, there is only one left, the St. Nicholas coal breaker. It is located midway between Mahanoy City and Shenandoah in the former village of St. Nicholas. The breaker location has often been cited as Gilberton, but Paul Coombe of the Mahanoy Historical Society says that is incorrect.

When locals noticed the demolition on the St. Nicholas in 2014, news reports falsely stated that the breaker would soon be history. Thankfully, as of this writing in 2017, the breaker still stands but is winnowing away as demolition continues.

The author was fortunate enough to be given entry into the breaker on two occasions by the Reading Anthracite Company, which still owns the structure. Security chief Richard Morgan was the guide to insure safety while demolition was ongoing. It was a kind gesture since the company had already been besieged by requests to photograph St. Nick, and trespassers were storming the site.

The fact that the St. Nicholas is such a massive building is probably what allowed it to outlast the region's other breakers. Dangerous materials like asbestos in its construction and its mammoth size make it expensive to demolish.

One of the most surprising revelations upon touring the facility is how well preserved it is. Despite missing windows, steel and concrete remain stubbornly intact, a testament to its solid engineering.

Coal breakers are unique to the coal region, which is comprised of Carbon, Columbia, Lackawanna, Luzerne, Northumberland, and Schuylkill Counties. Since anthracite coal must be broken down and sized, breakers are necessary to prepare the coal for sale. Bituminous coal does not require the same preparation, thus eliminating the need for a breaker. Breakers were a crucial part of anthracite's journey to market.

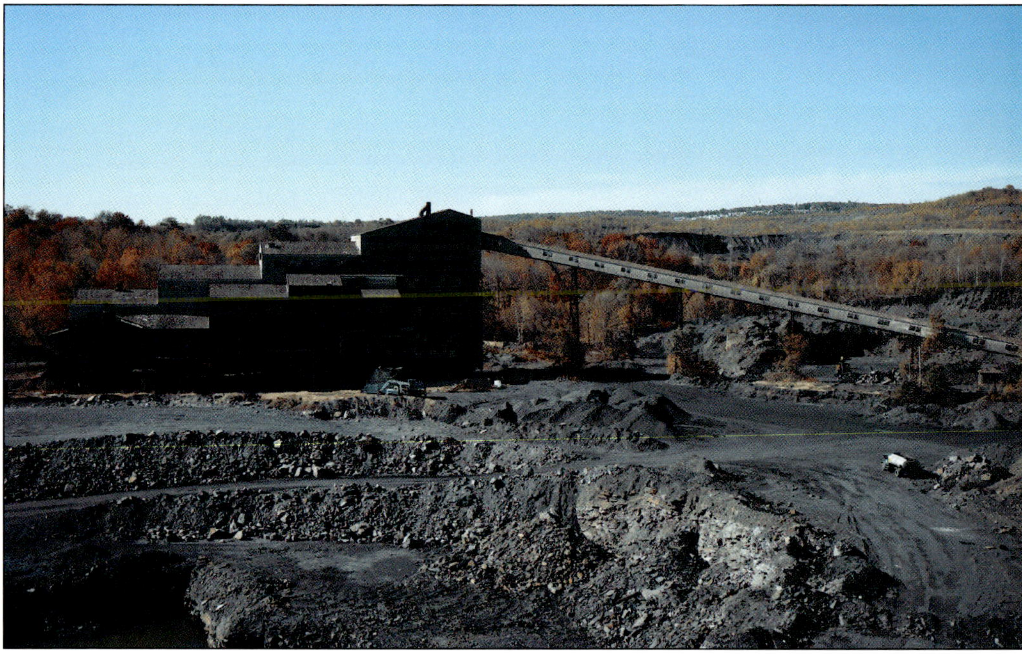

Demolition on the St. Nicholas began in 2014. Initially, outbuildings were the first to go. Now, the dismantling of the colliery continues, and it may take a while to complete since St. Nicholas was, at one time, the largest breaker in the world. In this photograph, the breaker and its coal conveyor remain intact.

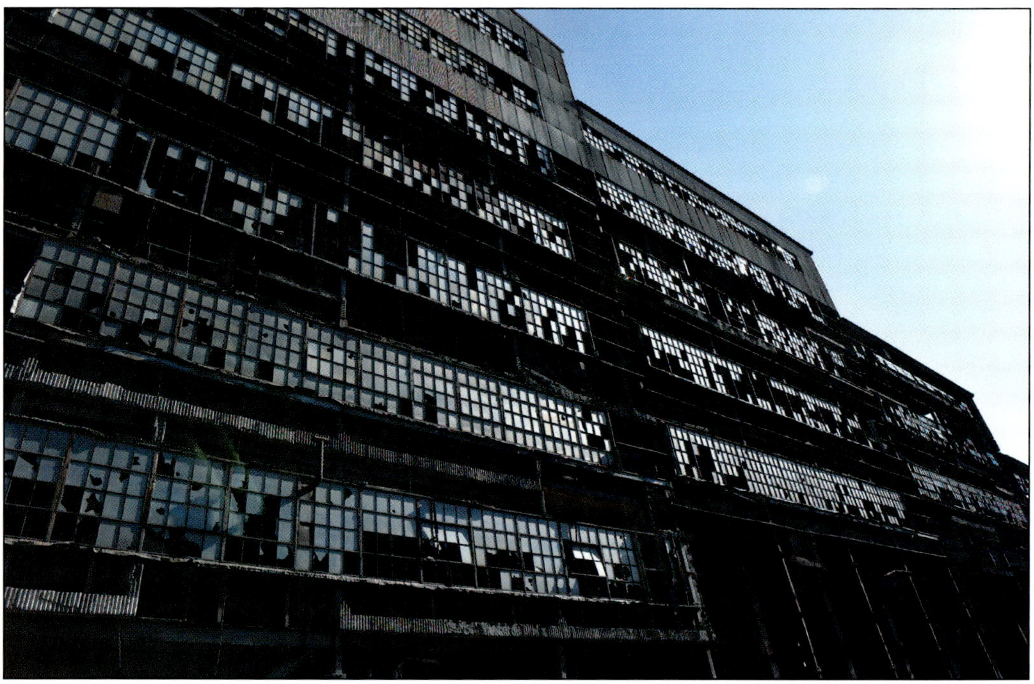

The breaker's job was to take coal from the mines and then process it. Processing included cleaning the coal of impurities that included rock and shale. It was then broken down into different sizes based on how it would be used in different types of furnaces and stoves.

The St. Nicholas's anthracite was crucial to the growth of the Industrial Revolution and, later, to winning World War II. Supey says, "The coal was used for battleships, home heating, schools, hospitals and government buildings."

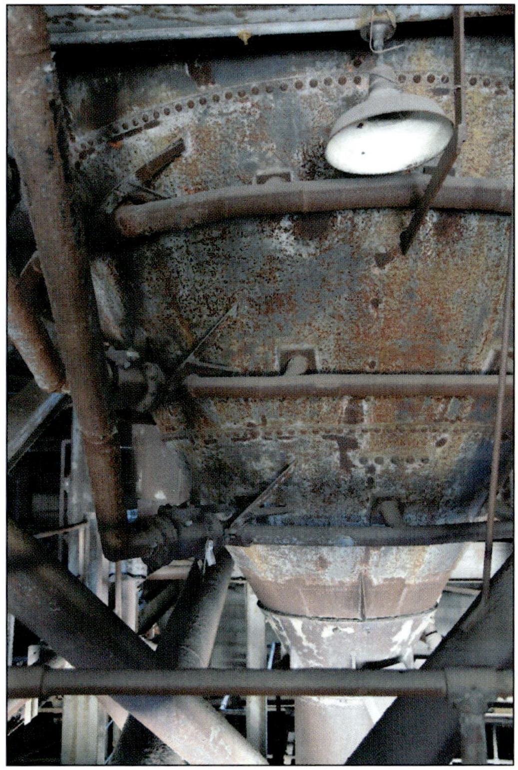

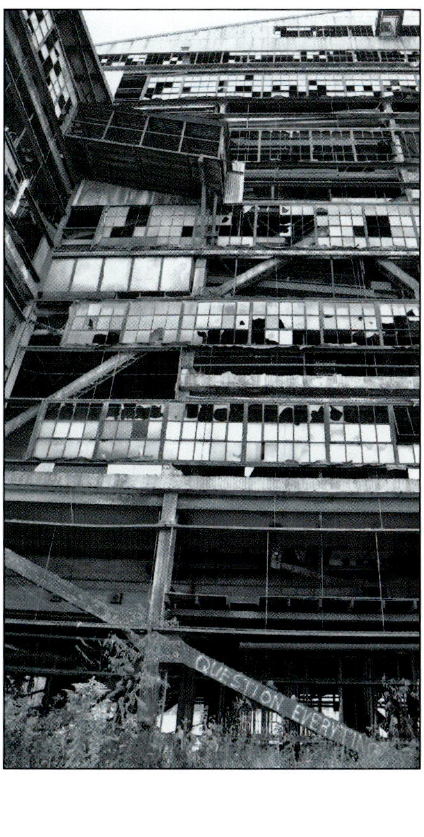

Like other coal breakers in the region, the St. Nicholas has become a canvas for graffiti artists. Throughout the colliery, there is one tagger who managed to leave his or her mark on almost every floor. The tagger simply sprayed "Flo." In this photograph, it is unclear what the tagger meant by "Question Everyting." One thing is clear, though—the tagger is not adept at spelling.

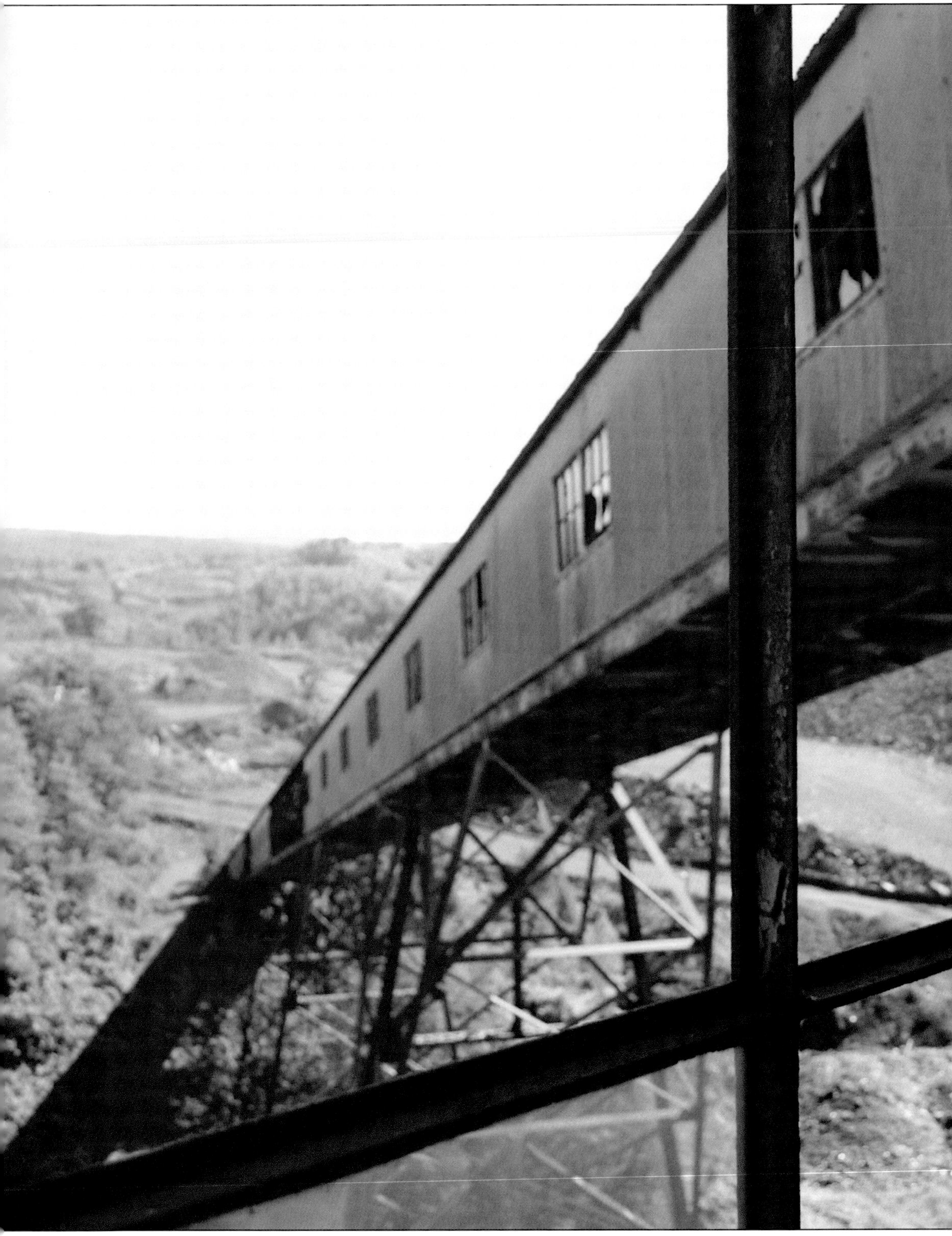

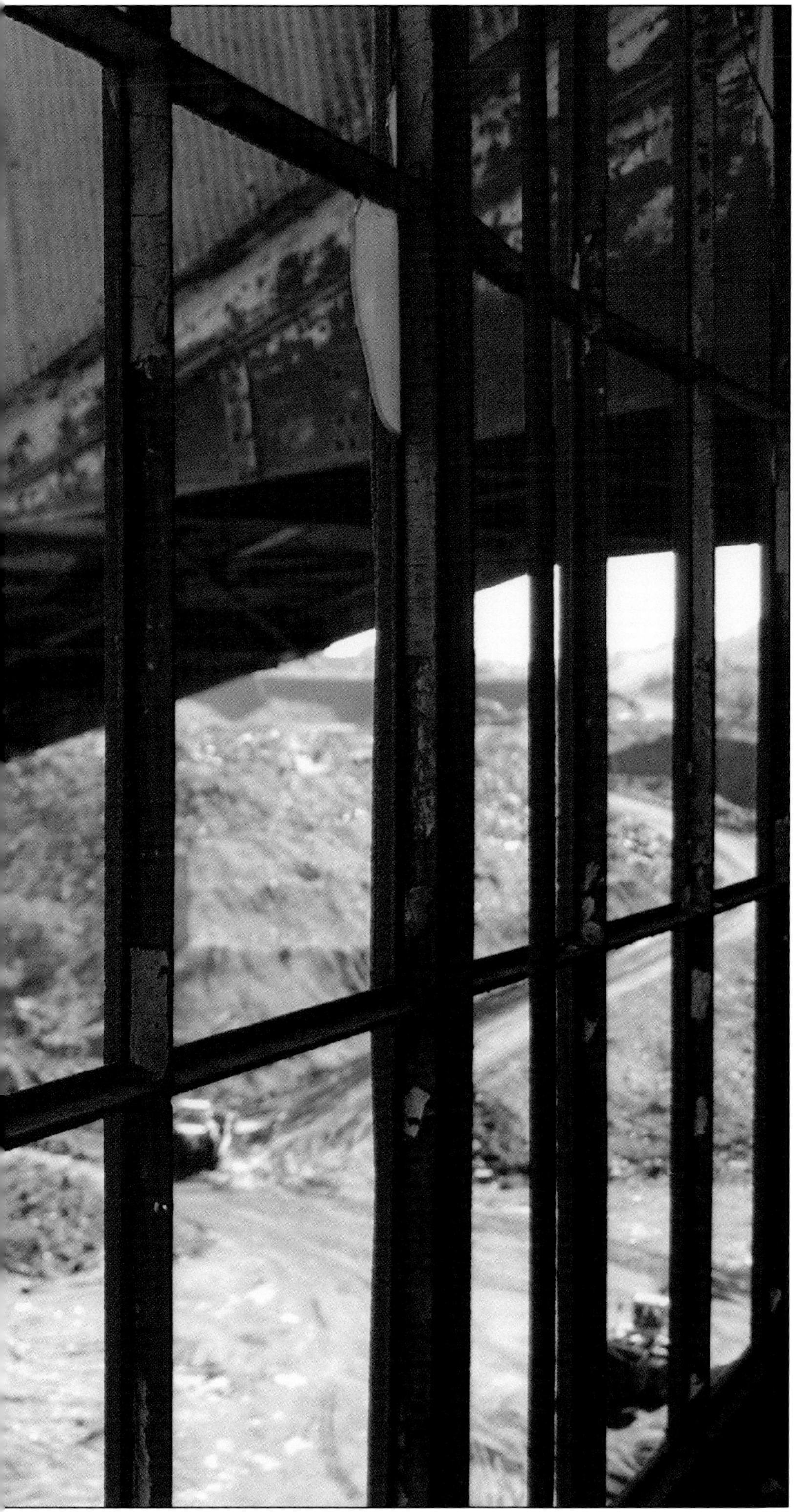

At one time, 12,000 tons of anthracite coal were processed each day. It took only 12 minutes for coal to complete its journey through the breaker so it could travel to market. Pictured here is the main conveyor belt that brought coal to the top floor of the breaker.

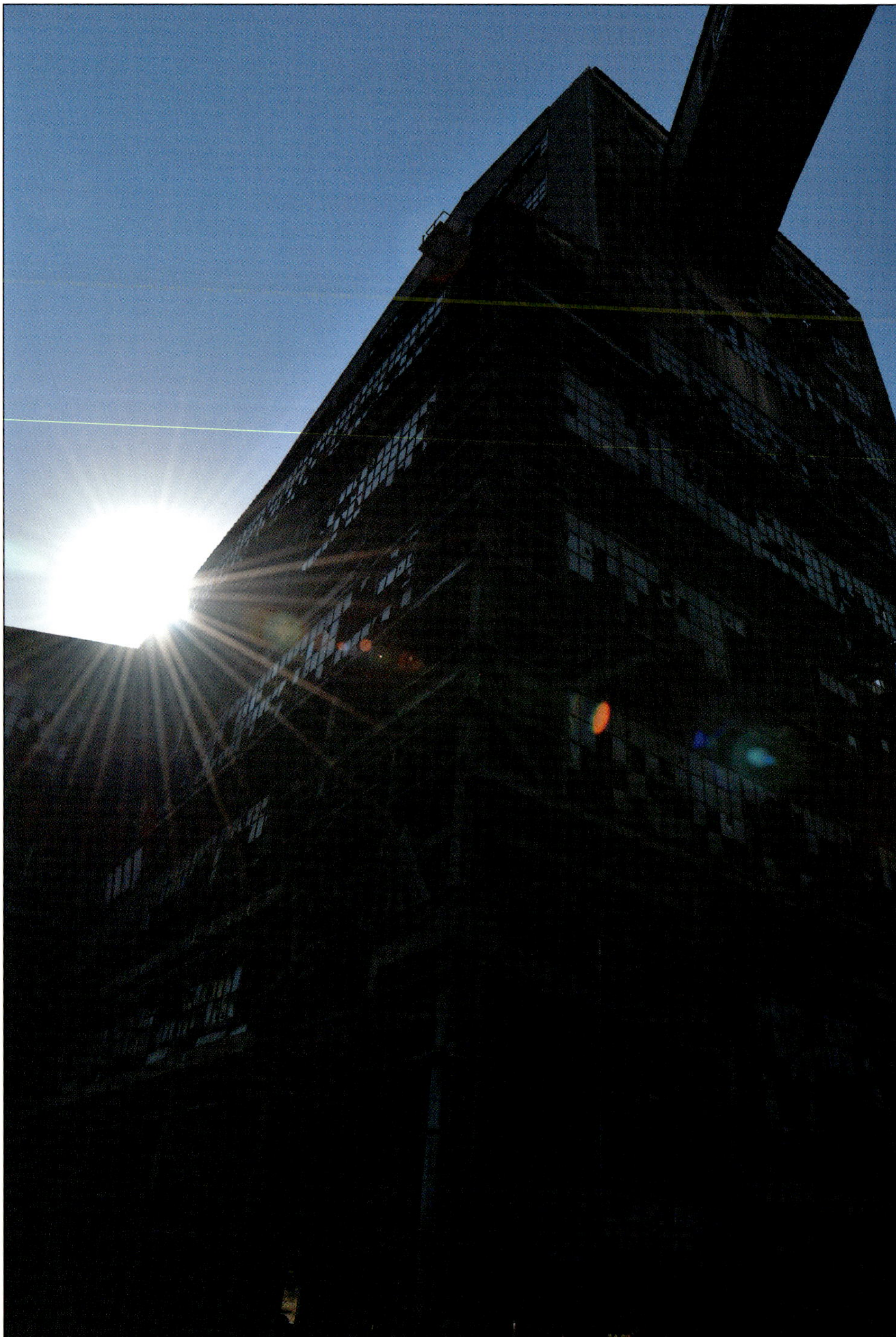

The St. Nicholas Breaker, seen in this photograph, is not the first breaker to grace this site. According to Coombe, "There were actually three St. Nicks. The former ones were made out of wood." The present structure started operations in 1931. Today, locals often refer to this breaker as Old St. Nick or even New St. Nick when, in fact, it is actually St. Nick III.

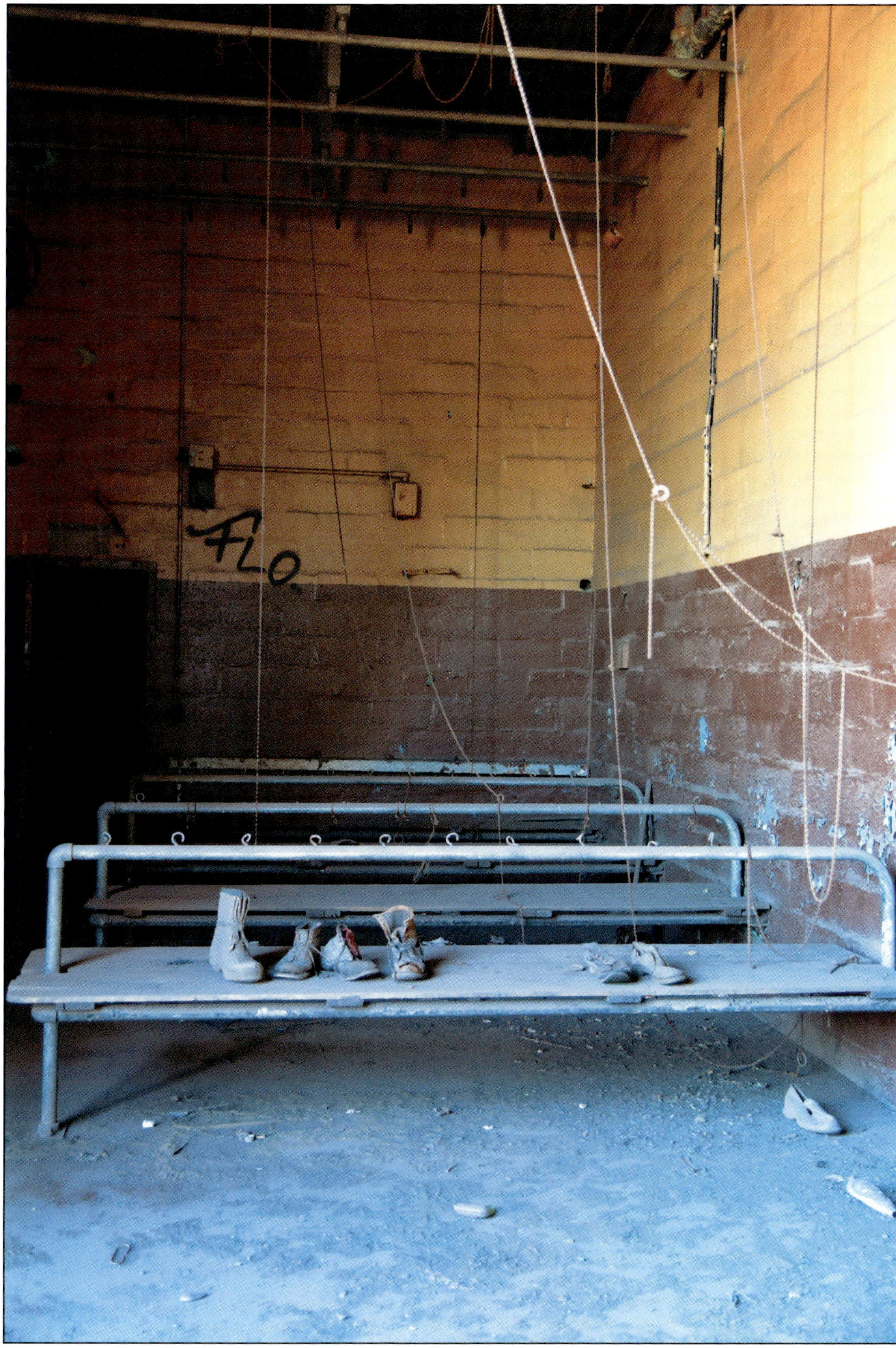

Much like a locker room, this shifting shanty served as a place for workers to change in and out of clothes. At the end of the workday, clothes were tethered to thin chains and hoisted to the ceiling where the warmer air would dry them. Since working in a breaker was a dirty job, workers chose not to wear work clothes home.

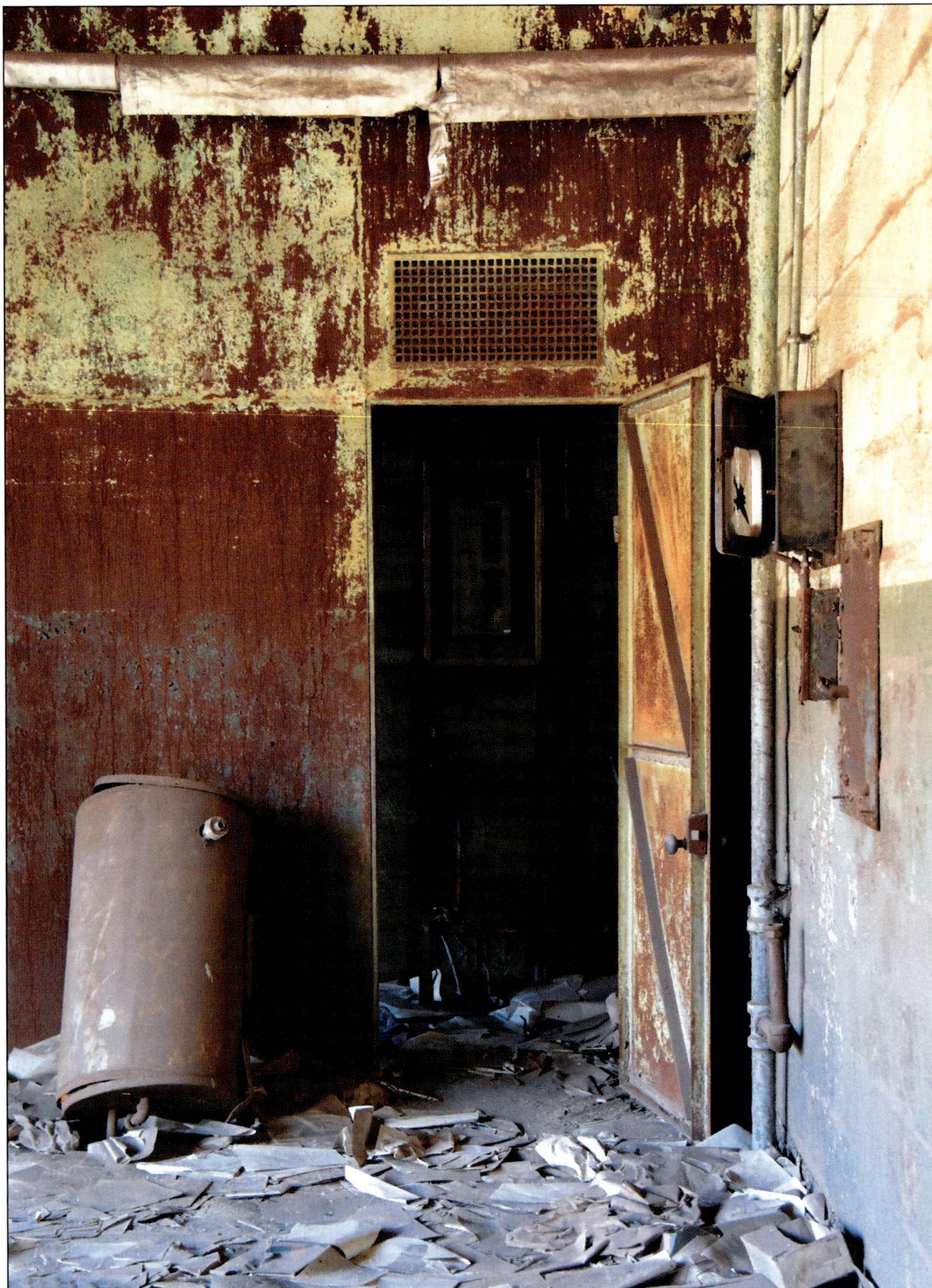

Those in management positions had the luxury of a private washroom. Additional perks of the job included higher pay, better housing, and education for their children. One of the downsides for management was intimidation and even violence wrought upon them by a secret Irish society called the Molly Maguires.

The St. Nicholas was something of a super breaker. Within the main structure, there were two independent lines that broke down the coal on opposite sides of the breaker. That insured work could still continue if one side became nonoperational.

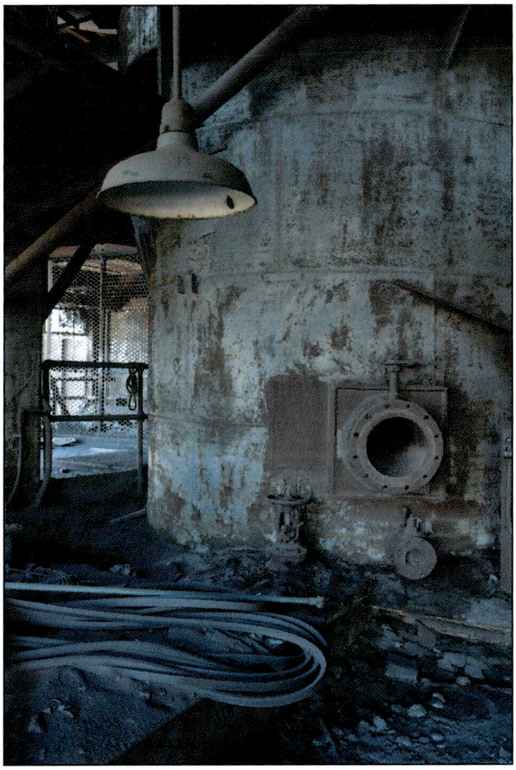

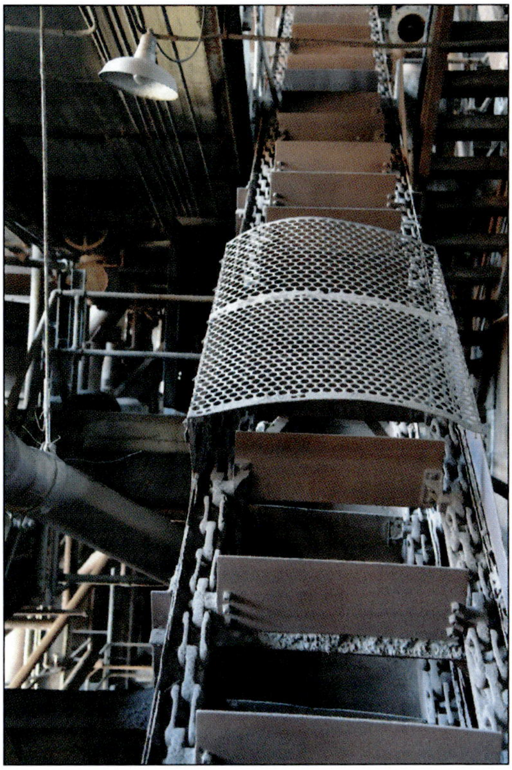

St. Nicholas was a central breaker. "Big hunks of coal didn't have to go too far [from the mine] to cut them into size for sale. As more railroads came in, they could take the coal and transport it to one breaker," says Coombe. As a result, smaller breakers were shut down.

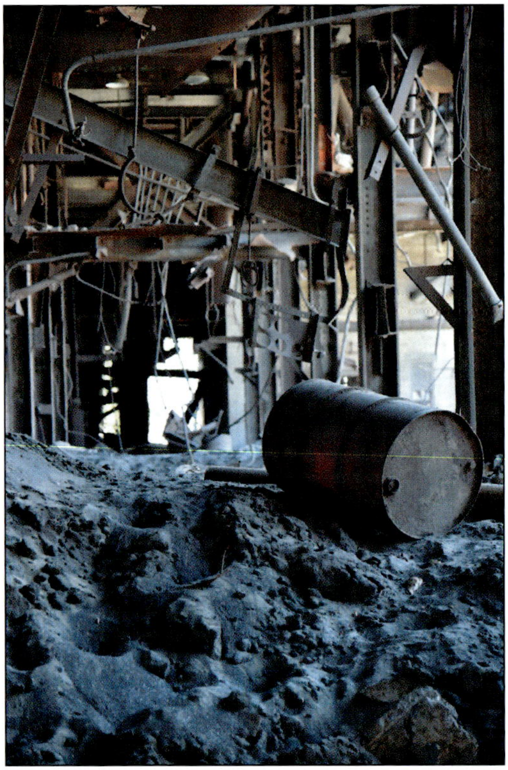

By the time the St. Nicholas was built, breaker boys were not working in breakers. Boys as young as eight years old worked 10-hour days, six days a week in the 1800s. "They weren't there working because they wanted to be. They had to be there because some families had 15 kids and the father didn't have enough money to put food on the table," says Supey.

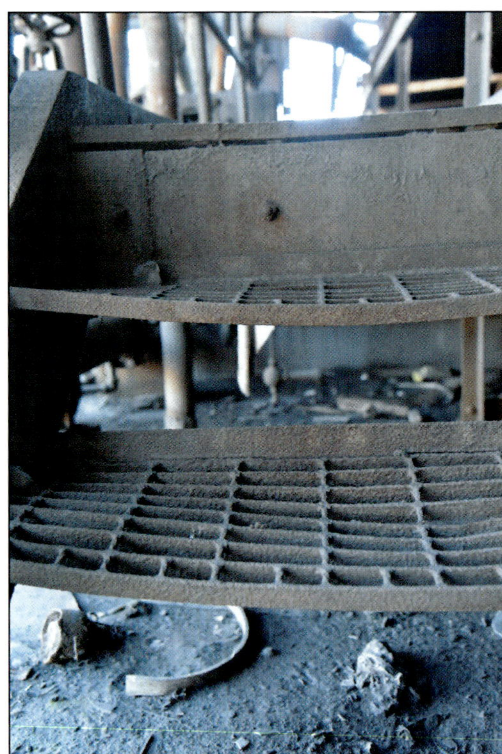

Dangers of the job included bloodied hands and even death from falls off catwalks. Because of technological innovation and child labor laws, the 20th century set free boys younger than 16 from breakers. Older men, or ones maimed in the mines, also toiled in breakers.

The collieries of Schuylkill County attracted workers from around the world. In the 19th century, immigrants from Wales, England, and Germany flooded into the industry. By the 20th century, they were joined by Polish, Lithuanian, Russian, Ukrainian, Slovakian, Serbian, and Italian ethnicities.

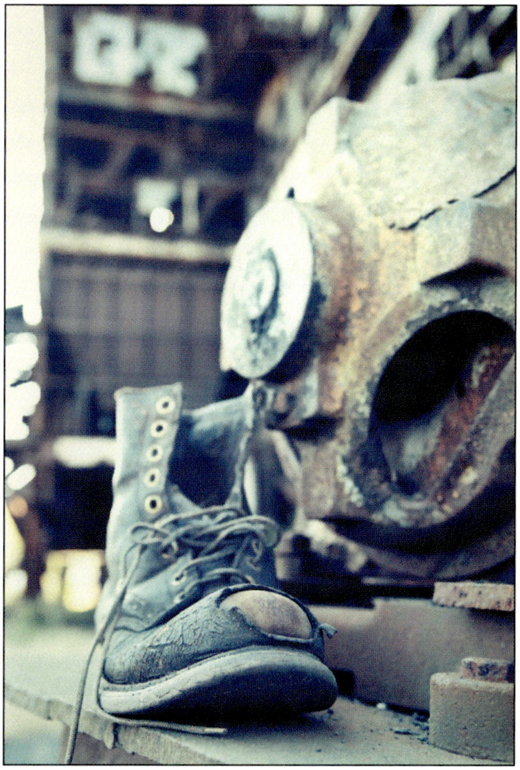

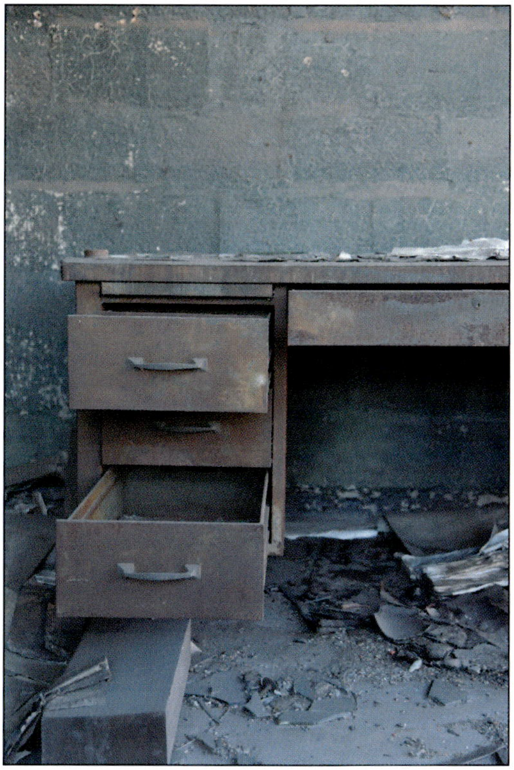

The majority of employees in the breaker did not have the convenience of a desk. Breaker workers toiled long hours on their feet for little pay. Supey says of those in the mining industry, "The only positive thing I can see [that came from that work] is that it gave people a really good work ethic. They say the work ethic from Northeastern Pennsylvania is second to none."

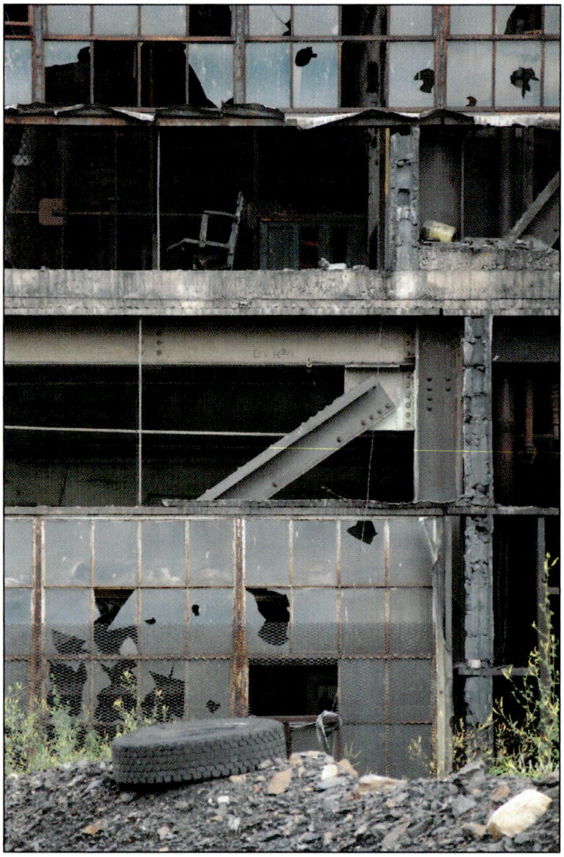

St. Nicholas got its name from one of the previous coal breakers on the site that was opened on Christmas Day. Coal was sold in red sacks that resembled Santa Claus's bag of toys. One of the coal patches that grew up around the breaker was also named St. Nicholas, but is no longer in existence.

According to Undergroundminers.com, the St. Nicholas contained 20 miles of pipes, 25 miles of conduits, 1.5 miles of conveyor lines, and 118 miles of wire. In addition, it had 3,800 cubic yards of steel. Any of those materials that were not already stolen from the breaker are now being scrapped during the demolition process.

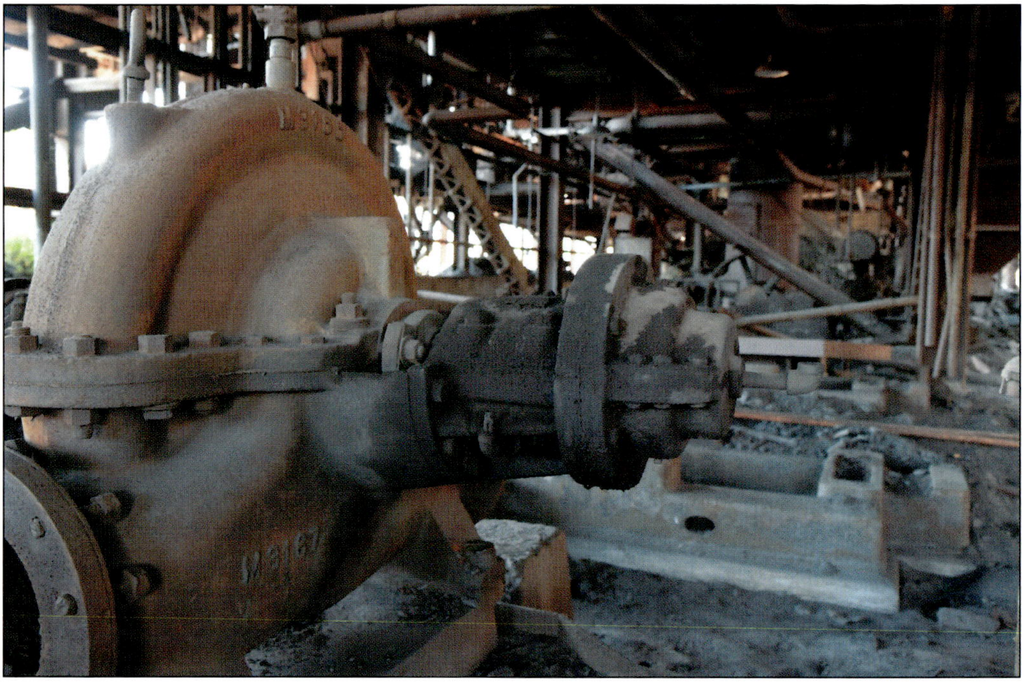

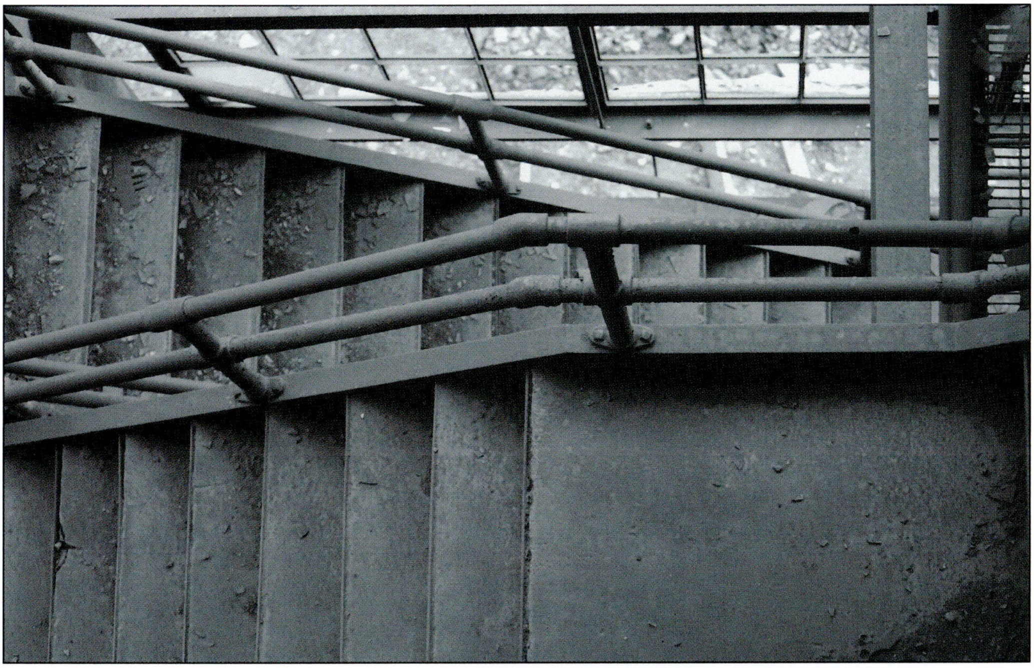

With 10,000 cubic yards of concrete used during construction, the St. Nicholas is the size of a city block. Unfortunately, pictures cannot fully convey the behemoth that it is.

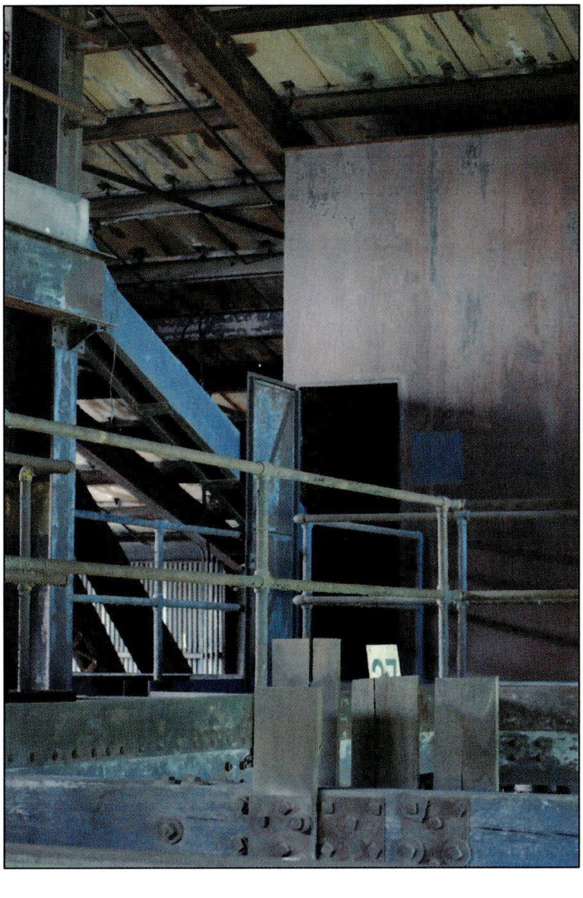

The Reading Anthracite Company, which was founded in 1871, still owns the St. Nicholas. It has been inundated with visitors to the site since it announced the commencement of demolition. It is private property, and anybody found within the building will be cited with trespassing. It is not only illegal to enter but also dangerous while demolition is ongoing.

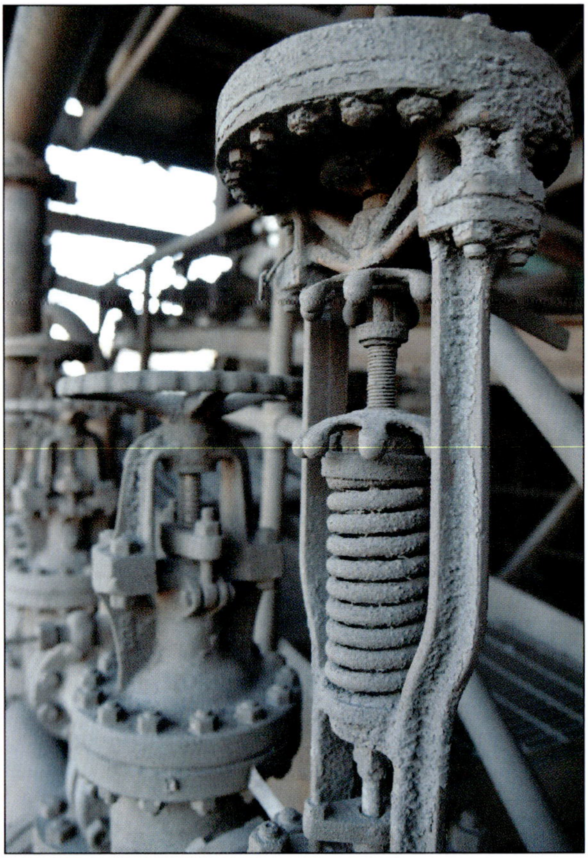

Coal breakers are also known as tipples or collieries. Anthracite needs to be broken down because it is crucial to have an even airflow around the coal so it will burn hotter and longer. In 1808, Wilkes-Barre tavern owner Jesse Fell was the first person to burn coal on an open grate. His discovery lit the public's desire to burn anthracite in homes.

It is remarkable that so much of this structure's essence still remains after being closed for almost 50 years. There were hopes to preserve the breaker as a museum, but funding never materialized to do so. Since the St. Nicholas is such a massive structure, it would have taken millions of dollars to restore it.

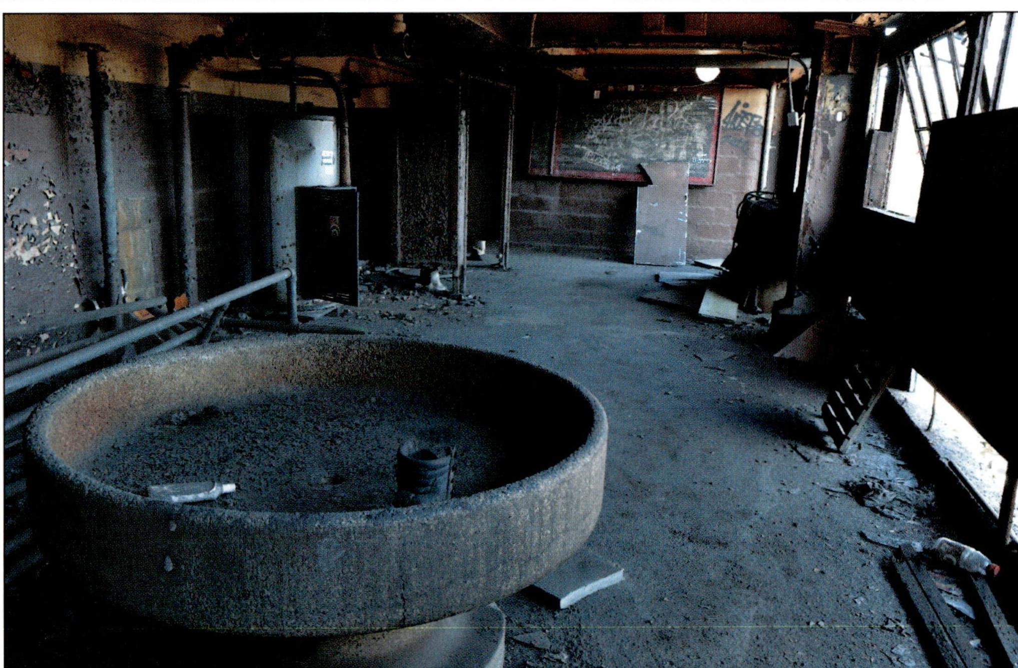

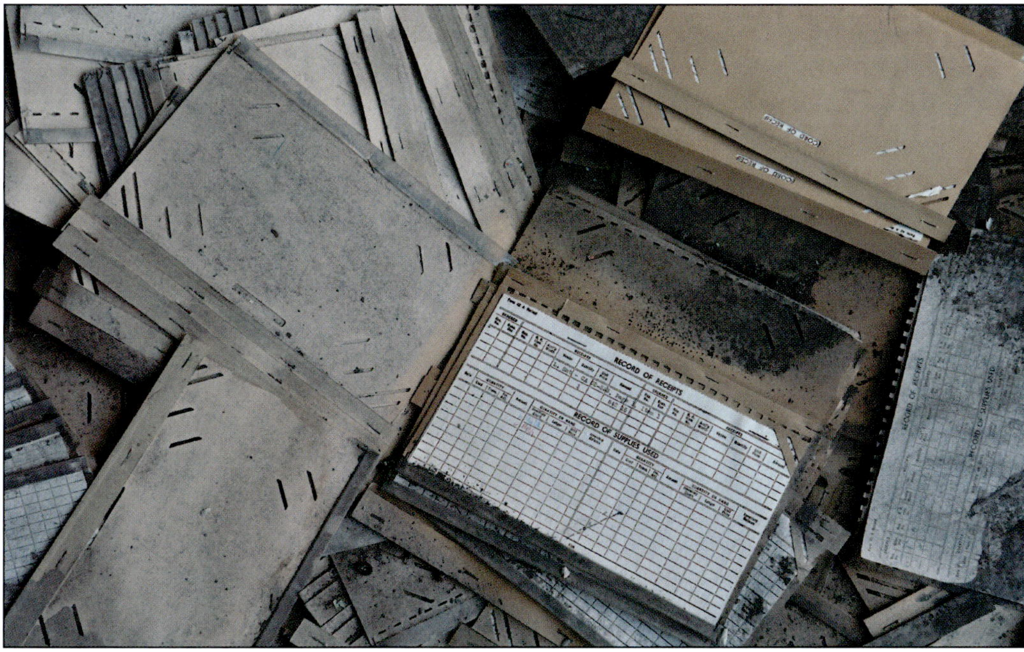

Hundreds, perhaps thousands, of office documents spill upon the floor. Centrally located in the photograph above, a record of receipts has two handwritten entries for inventory counts. The first was on July 10, 1956, and the second was on August 9, 1957. It is remarkable these papers have not been destroyed after decades in a windowless room.

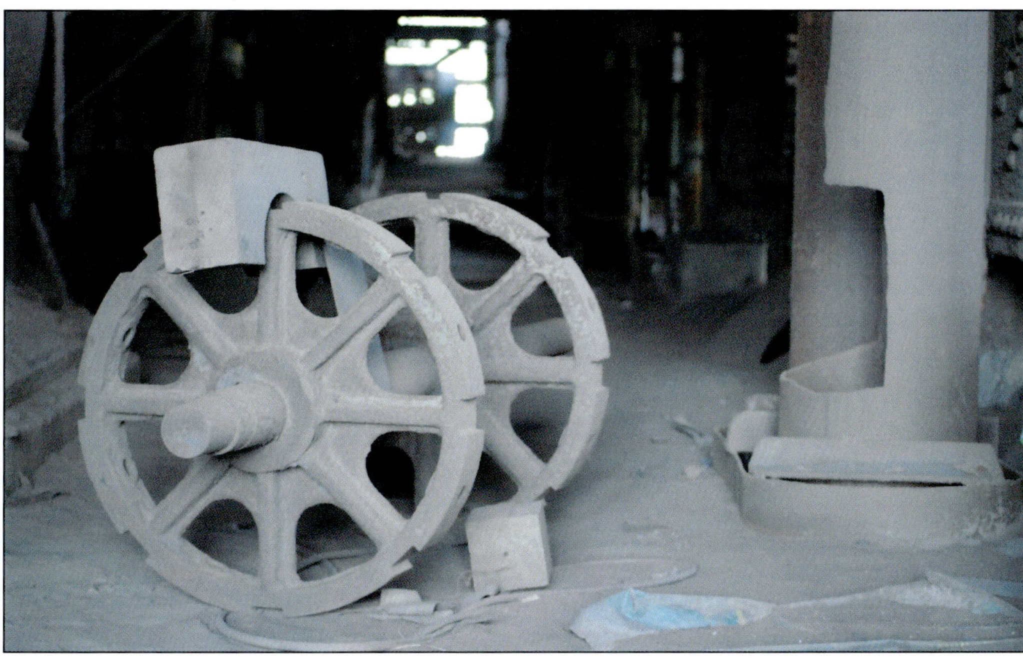

The St. Nicholas Breaker was shuttered in 1965, long after the heyday of coal. It was once one of 42 collieries in the Mahanoy Valley, according to the Mahanoy Historical Society. Throughout Northeastern Pennsylvania's coal region, it was one of hundreds. It now stands as the sole survivor of a bygone era.

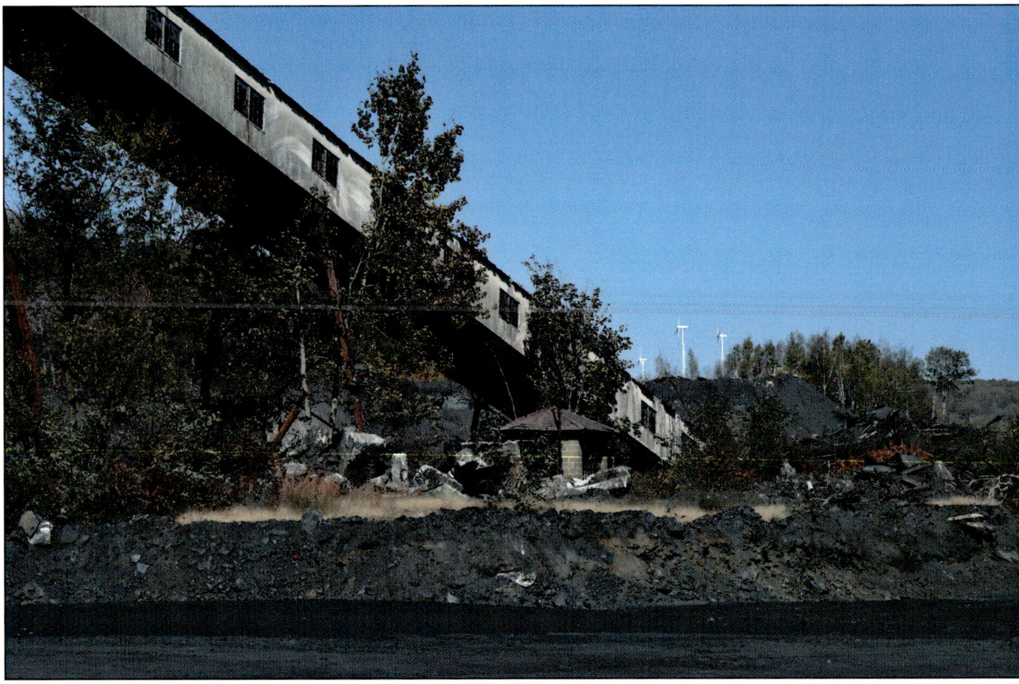

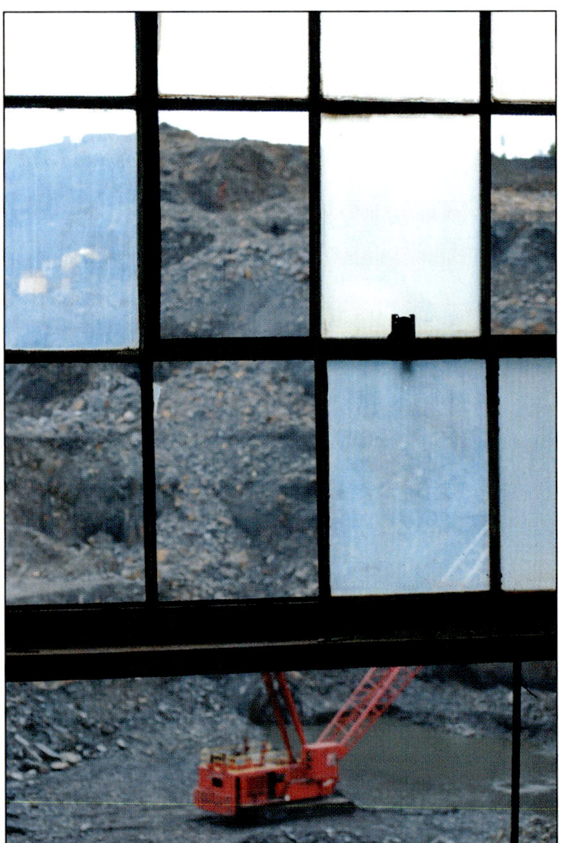

In the mountains beyond the conveyor belt, wind-powered turbines can be seen. Their lurking presence almost seems to mock the breaker's demise. Their shiny white spires reaching to the sky are in sharp contrast to the conveyor that has been buried into the black ground.

Through the window of the St. Nicholas, the present state of coal mining is revealed. Rather than men toiling in mines, big equipment now shoulders the burden of work. According to the US Energy Information Administration, in 2014, there were 11 underground anthracite mines left in Pennsylvania while there were 53 strip-mining operations. Breakers are no longer needed.

Two

Huber Is History

The Huber Breaker in Ashley served as a place of curiosity for locals for decades before its demolition. William C. Best, who is with the Huber Breaker Preservation Society (HBPS), often conducted tours there and encouraged visitors to take souvenirs with them. He says, "I thought it was my job to introduce people to it. I remember when they were tearing down the main conveyor line, you couldn't help but be saddened because people were not going to be able to enjoy what I did and enjoy our heritage. The breaker was a symbol of what had gone on for over a century and a half underground. It was visible."

That visibility made the Huber a landmark that could be seen for miles. Best says, "It was a welcoming site. When you were coming in from 81 north, you would see the colliery complex. It stuck out like a sore thumb. As a kid, I remember smoke coming out in the winter and at night it was still visible with lights around the rail yard."

The Huber was Wyoming Valley's last breaker, and with the structure's demolition, it was a sad passing for many in the Wyoming Valley. Another piece of history is gone.

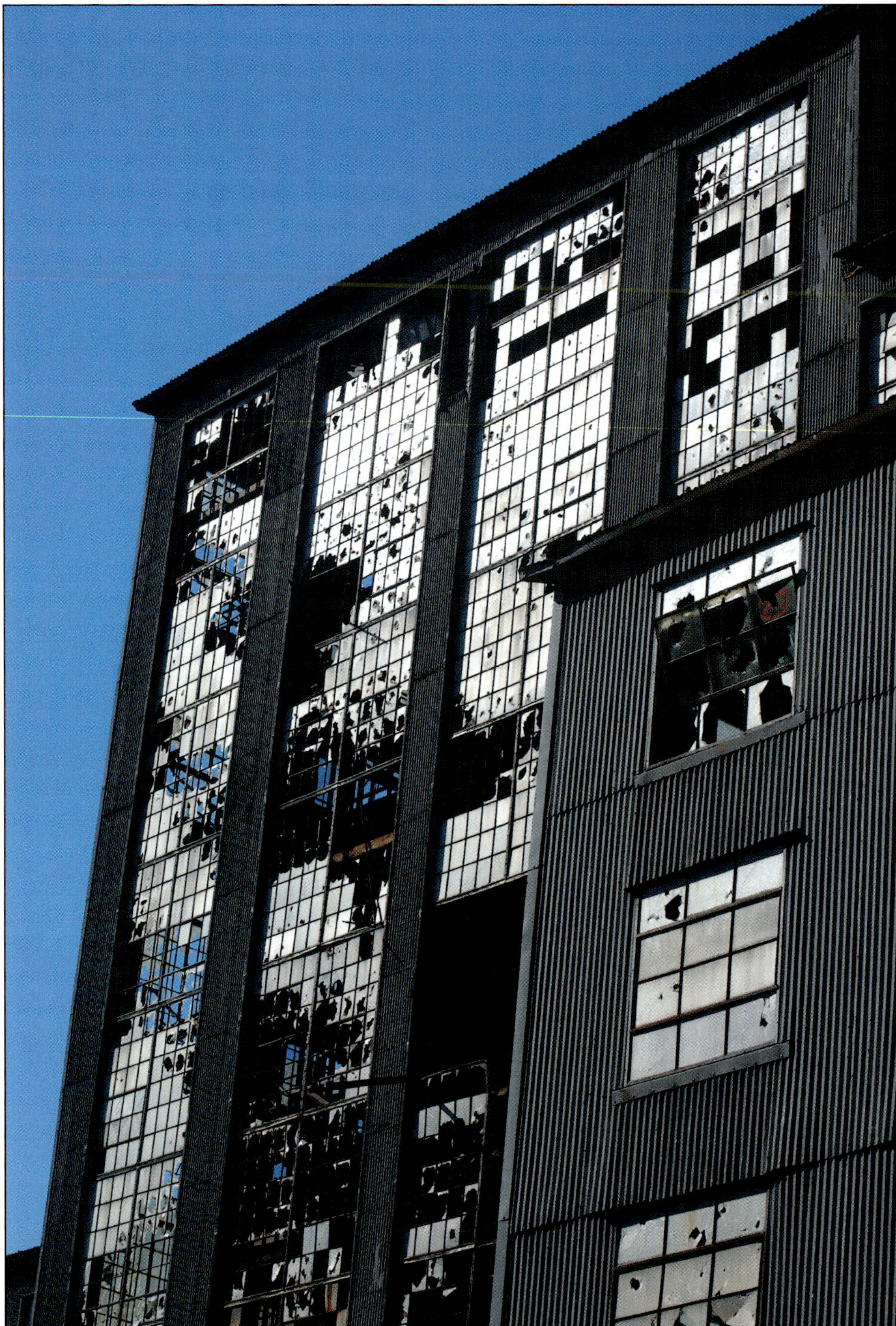

The Huber Breaker in Ashley, just outside of Wilkes-Barre in Luzerne County, is pictured here just months before its demolition in 2014. It was the second-to-last coal breaker to survive the golden age of anthracite in Northeastern Pennsylvania. Like the St. Nicholas Breaker, it once held the distinction of being the largest and most technologically advanced breaker in the world.

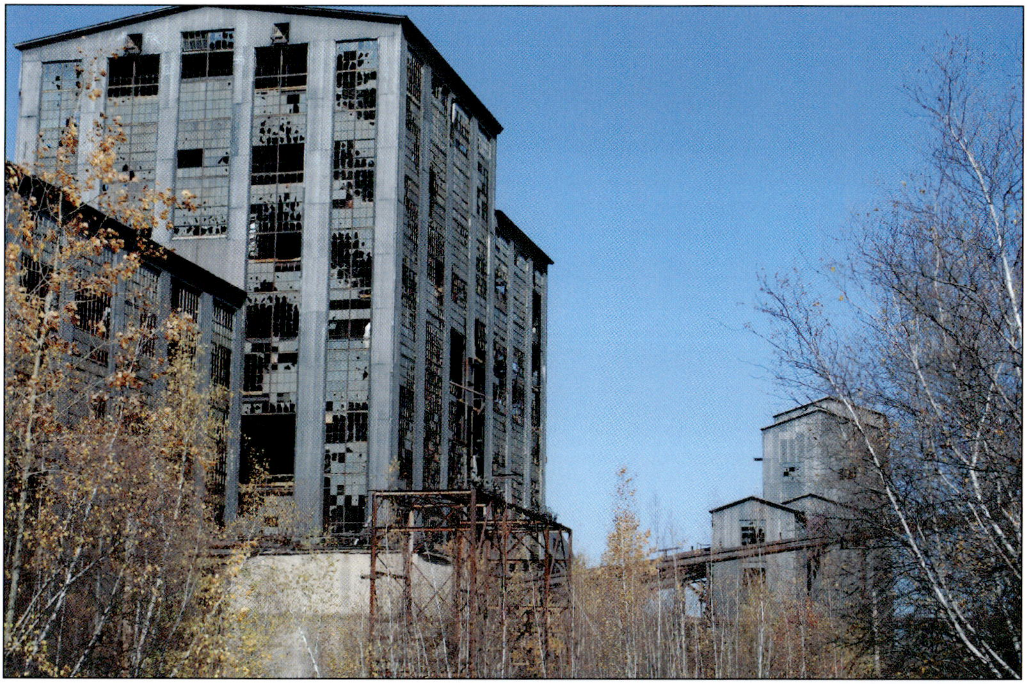

Pictured here is the Huber, which was the third breaker to occupy the site according to Huberbreaker.org. The first was built in 1855 by the Hartford Colliery company until a fire razed it in 1884. In 1895, the Maxwell No. 20 was completed and, at its peak, employed 1,755 people. In 1937, it too succumbed to fire.

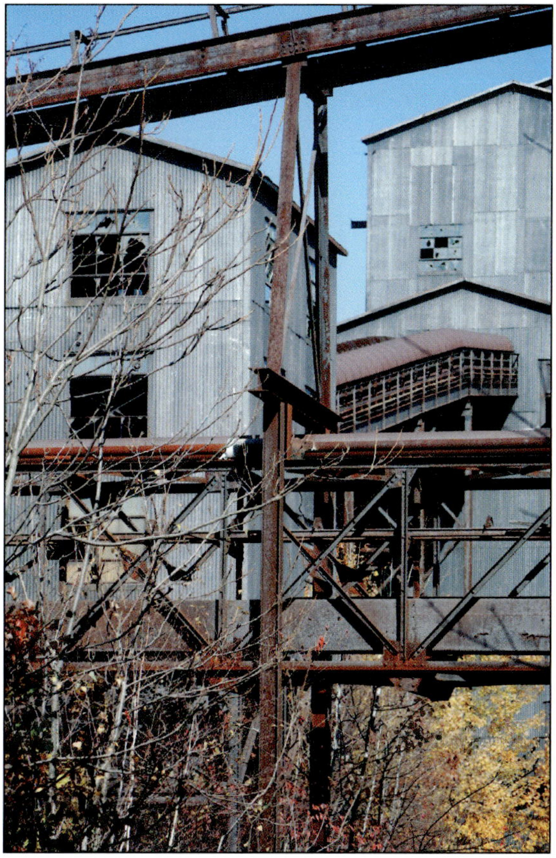

The Huber was constructed nearby in 1938 on the same site as the Maxwell No. 20. Unlike its predecessors, it was constructed of metal and concrete. The breaker and its surrounding buildings occupied 7¼ acres of the 26 acres the site encompassed. The 11-story breaker was close to 23,000 square feet.

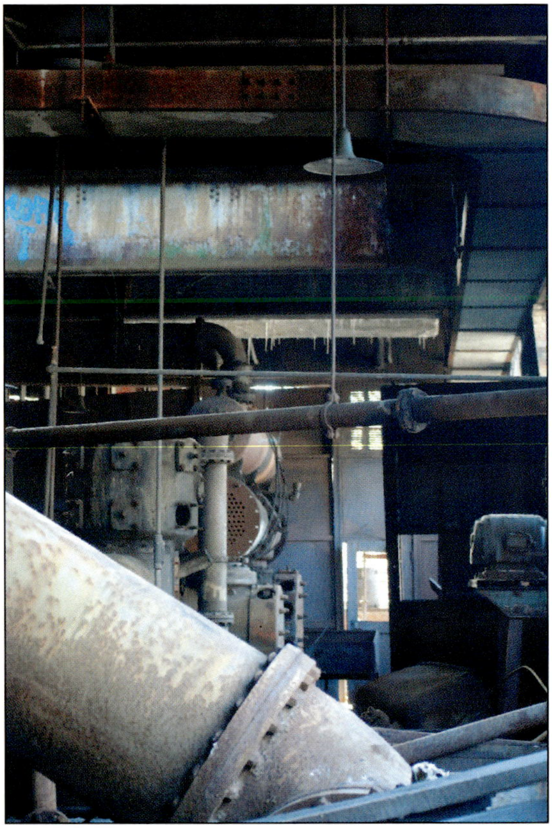

This interior view of the Huber shows the industrial intricacies of the structure. Production at the Huber peaked in 1942 with close to a million tons of coal brought to market by 1,503 employees. Over 800,000 tons were shipped by train, while 237 tons were shipped by truck. The remaining coal was stockpiled on-site for local markets.

The windows of the Huber were designed for functionality and safety. Ray Clarke with the HBPS says, "They built them so if rocks or explosions would happen, it would only crack the window rather than blow it out. They put so many windows in there so they would have more light and they didn't have to use electricity."

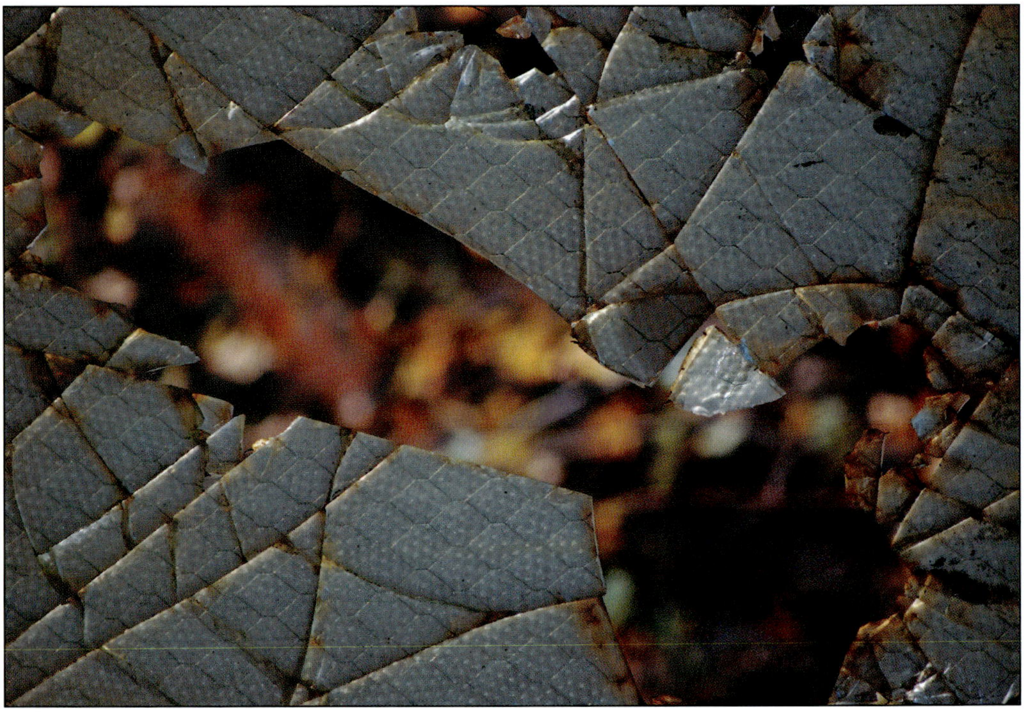

The Huber was an economic boon for Ashley. "It made the town what it was. When the breaker was open, there were shops, two dentists, six doctors, and three supermarkets in Ashley," says Clarke. The borough's peak population was 6,371 in 1940. According to the 2010 US census, the population was 2,790.

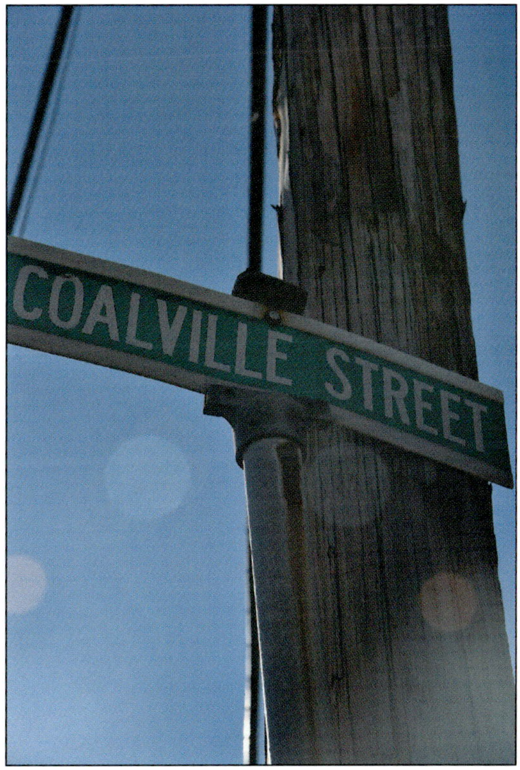

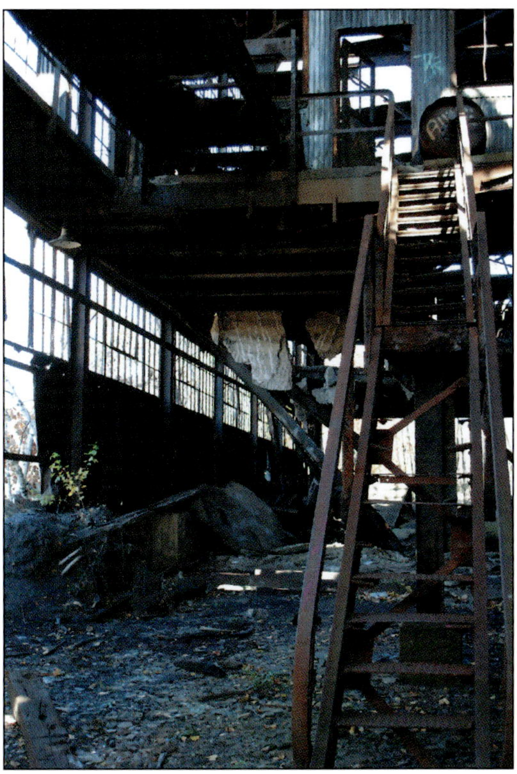

Clarke is one of the few employees alive to speak about the breaker. He worked there for only one year, hauling coal from the site. Even before employment, the Huber greeted him daily. "I looked at that building every day as a kid as I walked to school. It was part of my life. It was a big part of so many people's lives," he says.

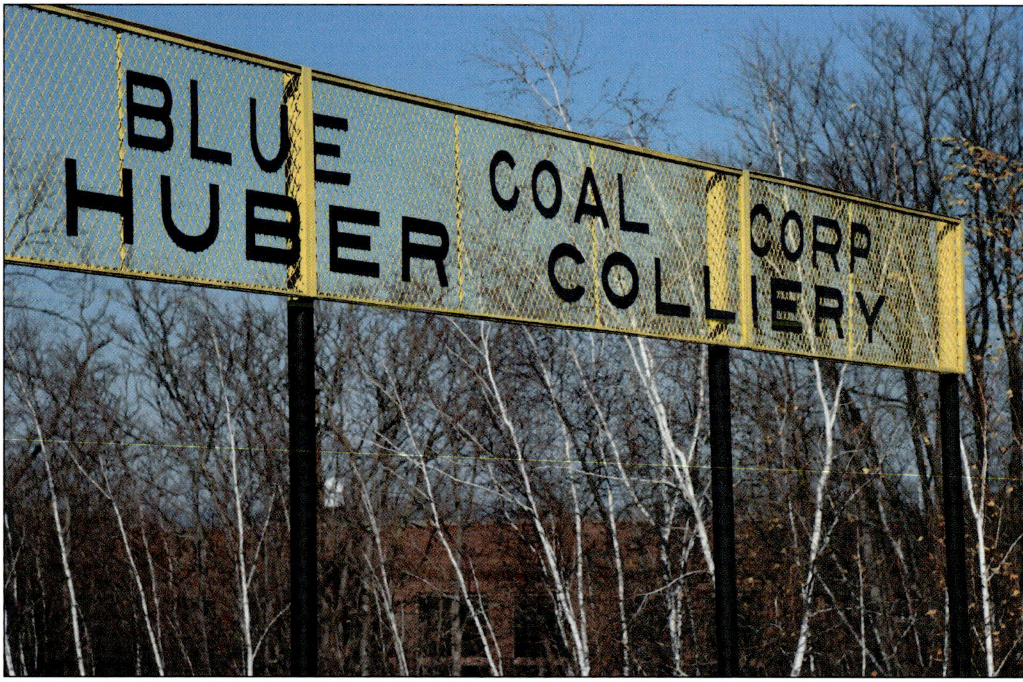

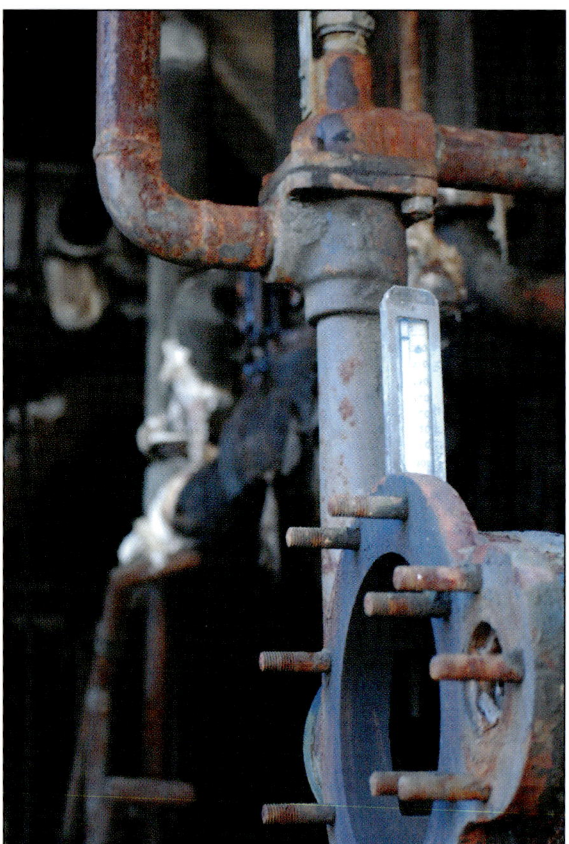

In the 1920s, the Glen Alden Coal Company developed a marketing strategy that employed dying its coal blue. This simply distinguished it from other coal to make it stand out from the black diamonds of other coal companies.

The colliery was capable of producing a ton of coal an hour that was used solely for home heating. "By the time it was built, there was no use for industrial coal," says Best. The anthracite coal was sent to the Eastern Seaboard and Canada.

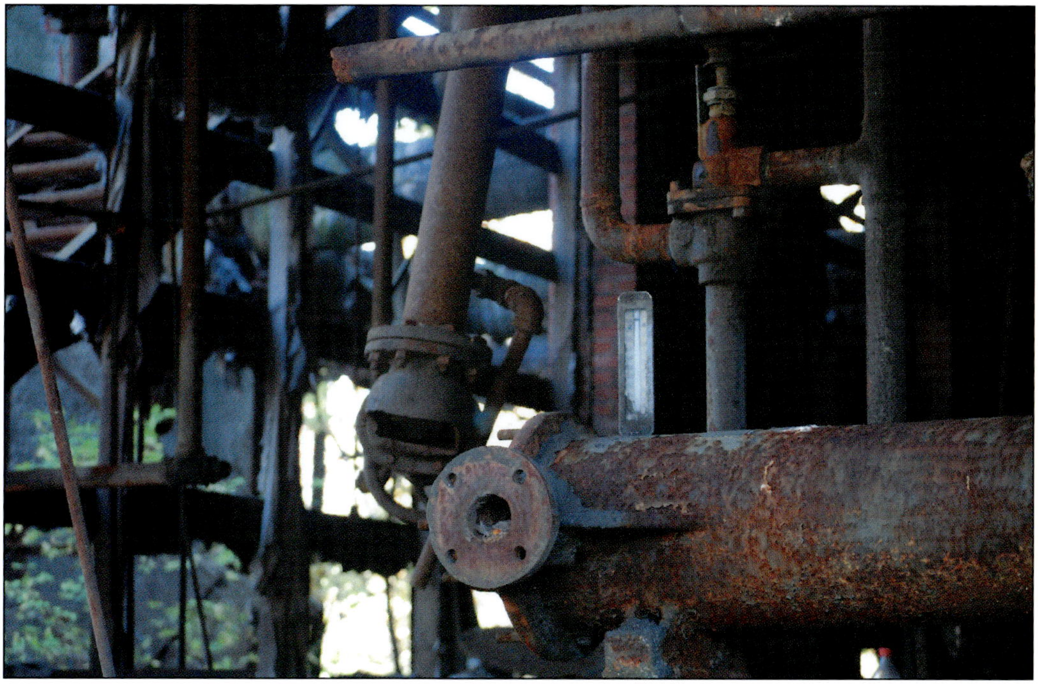

According to Huberbreaker.org, at the height of the Huber's coal production, 1,503 workers processed 979,208 tons of anthracite coal. Close to 100,000 tons were shipped by train and 237 tons by truck. The remaining coal was stockpiled and sold locally.

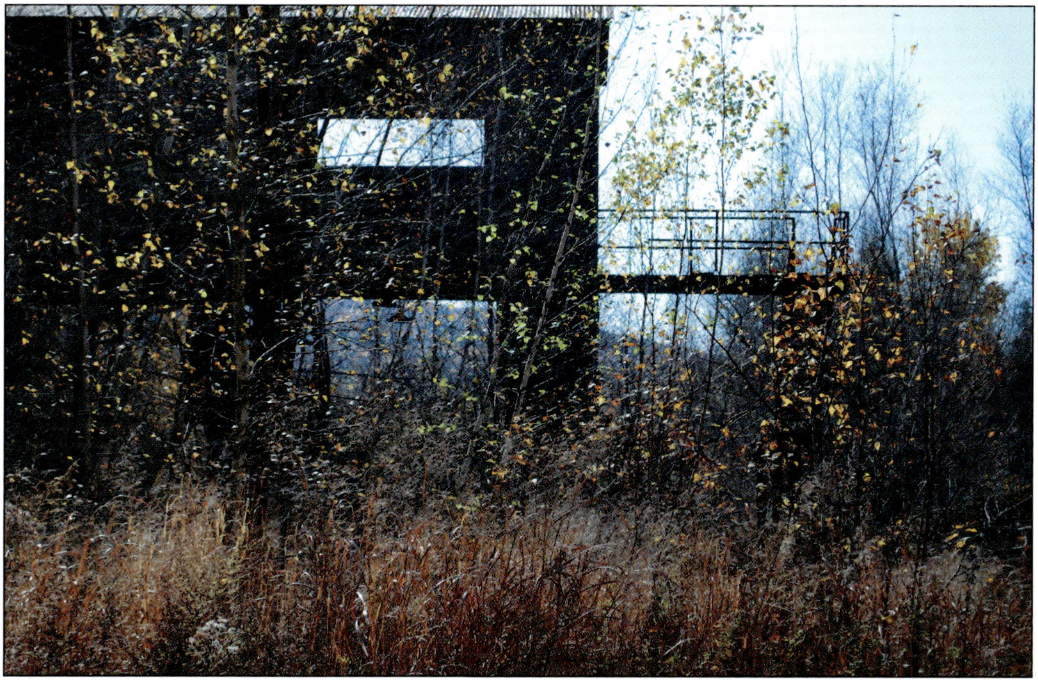

Pictured here is the Huber's coal inspector's station. The coal inspector's job was to randomly inspect every ton of coal to determine ash content and Btus (British thermal units). Within the on-site laboratory, results of the product were recorded.

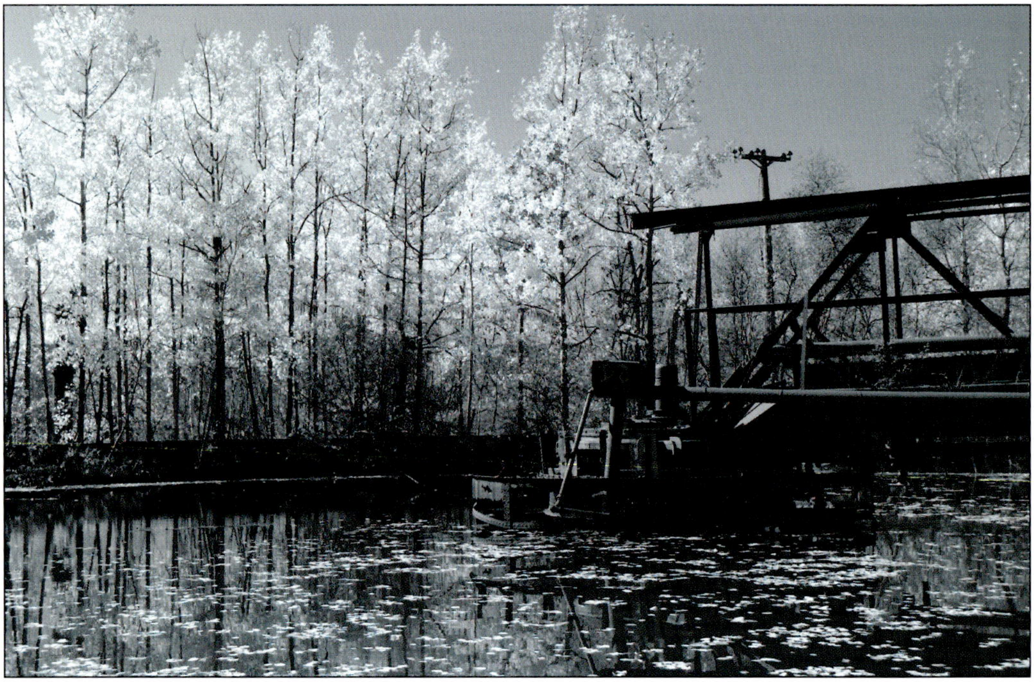

Located to the southeast of the breaker, the Dorr thickener used dirty water to cleanse smaller pieces of anthracite. Best says, "There were paddles at the bottom that would pick up the mud and pull silt back out to be processed."

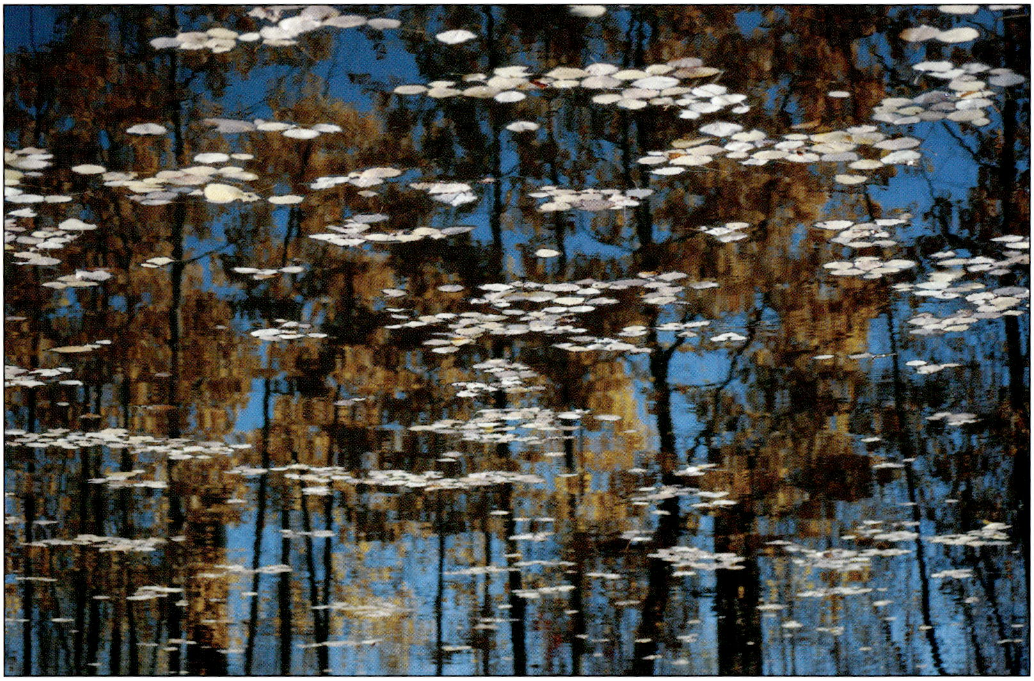

Pictured here is a reflection in the Dorr thickener's pool, which was 120 feet in diameter. Once waste settled to the bottom of the apparatus, water was then sent back for use in the breaker. Dorr thickeners are still used today in other industries across the world.

The Glen Alden Coal Company sold Blue Coal Corporation to Raymond Colliery for $6 million in 1966. Despite being one of the largest coal producers in the United States, Raymond experienced financial issues and sold the colliery to Raymond president James Durkin in 1972. The colliery was at the beginning of its end.

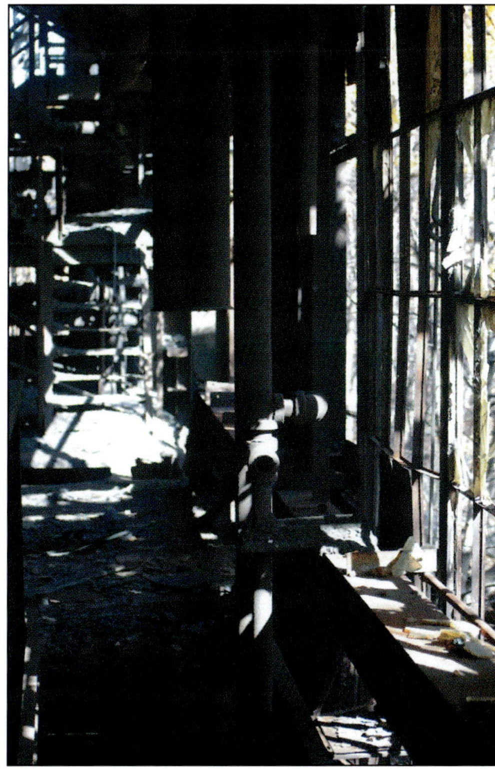

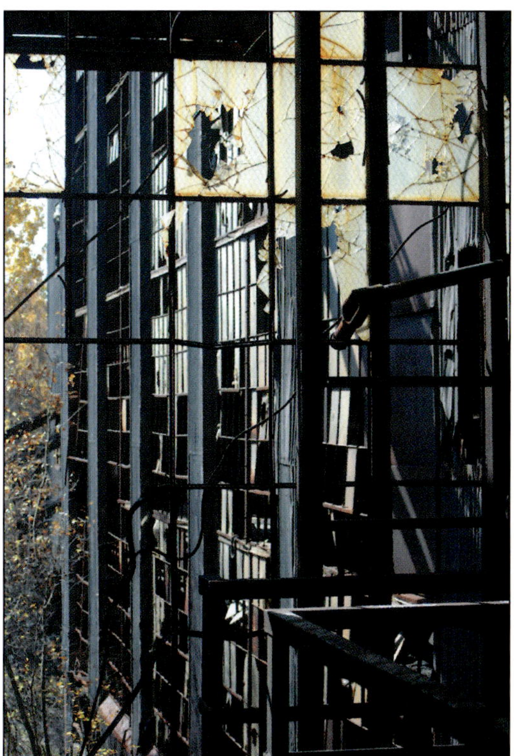

According to a Historic American Engineering Record report, one of America's largest crime figures was part of the Huber's history. The report reveals, "Durkin had difficulty securing the necessary finances to consummate the deal. He sought assistance from the Central States Pension Fund of the International Brotherhood of Teamsters and other sources, creating the Great American Corporation. . . . Fifty percent of Great American was actually owned by James R. Hoffa, Jr., who remained a secret partner until his disappearance."

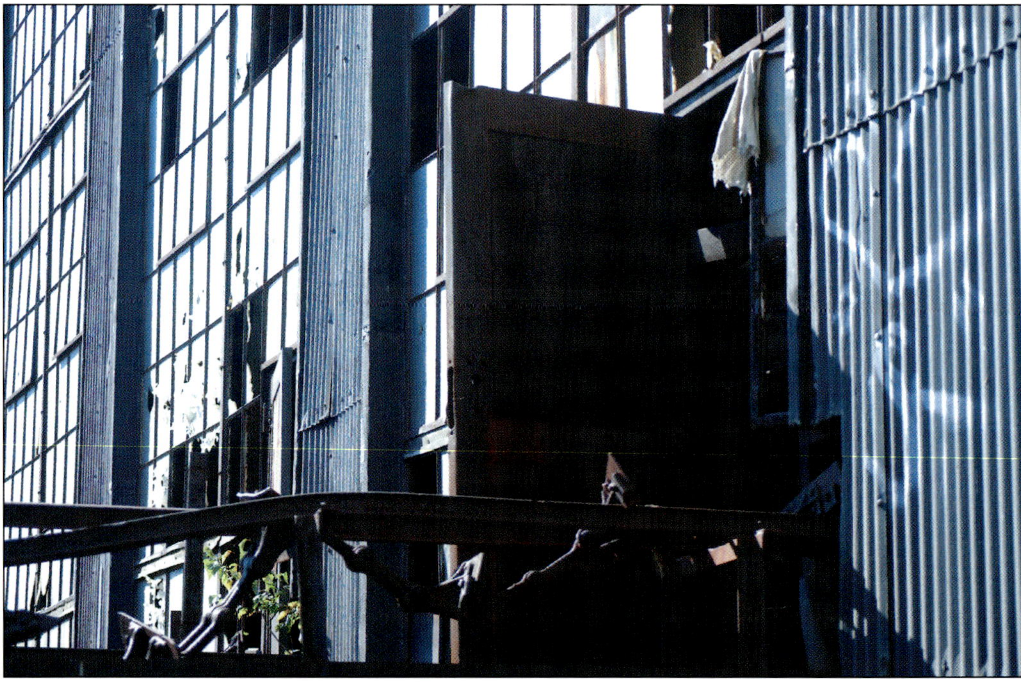

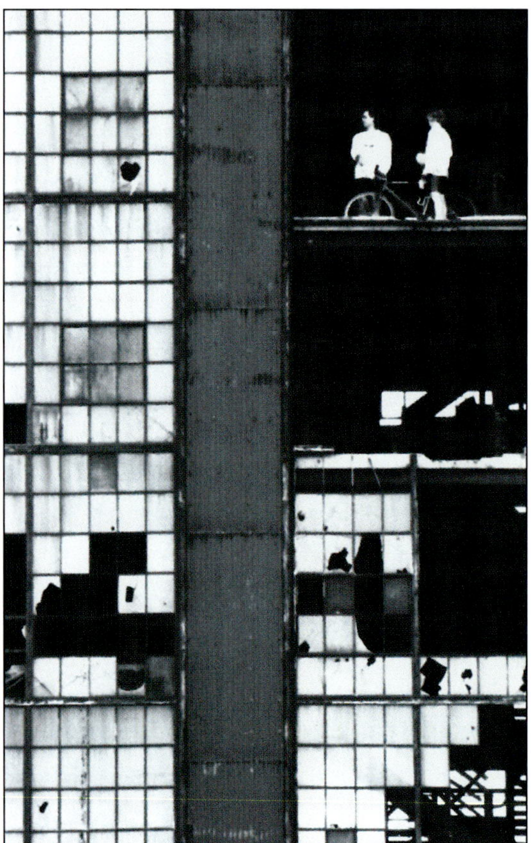

There were more attempts to keep the Huber operating, but all failed ultimately. Finally, in 1976, the Huber was shuttered for good. It, like the nearby mines, had become a casualty of changing energy needs in the country and worldwide. Oil and natural gas began dominating the fuel market, lessening the demand for anthracite.

The Huber was the last surviving coal breaker of the northern coalfield in Northeastern Pennsylvania. The abandoned behemoth started to decay and became a place for teenagers to congregate. It was also a magnet for photographers who wanted to document this relic of coal's heyday.

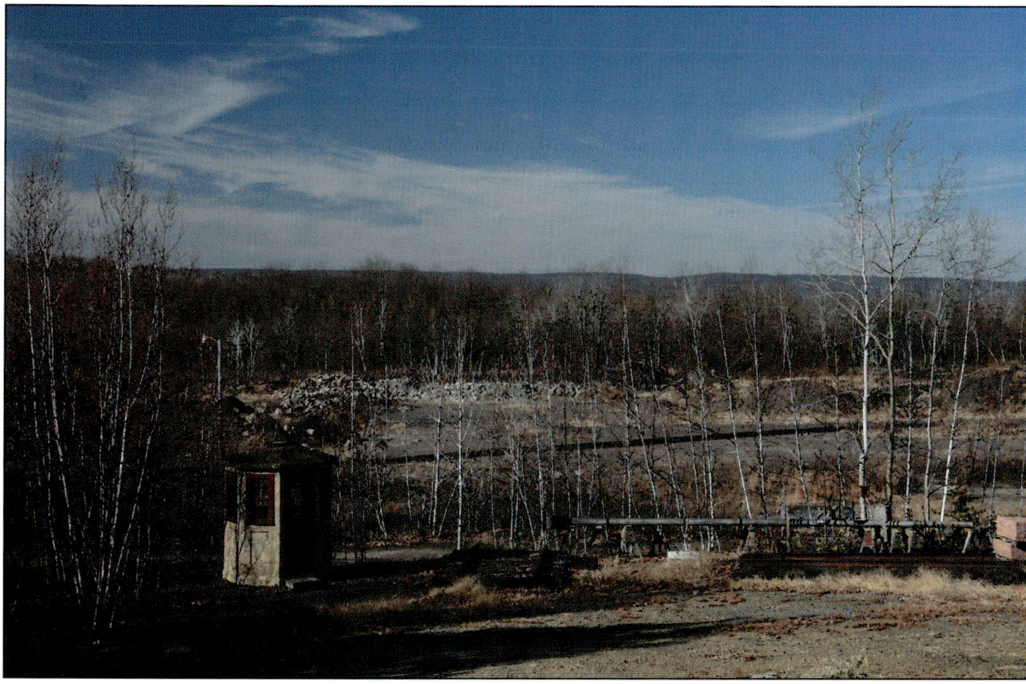

Thirty-eight years after the Huber's closure, Paselo Logistics, LLC, won a bid in bankruptcy court to purchase the breaker and its land. Initial demolition started January 24, 2014, with the breaker coming down on April 24 that year. Clarke witnessed the demolition saying, "They pulled it down with cables. They cut the columns with torches and then put cables around the base and pulled it down with heavy equipment."

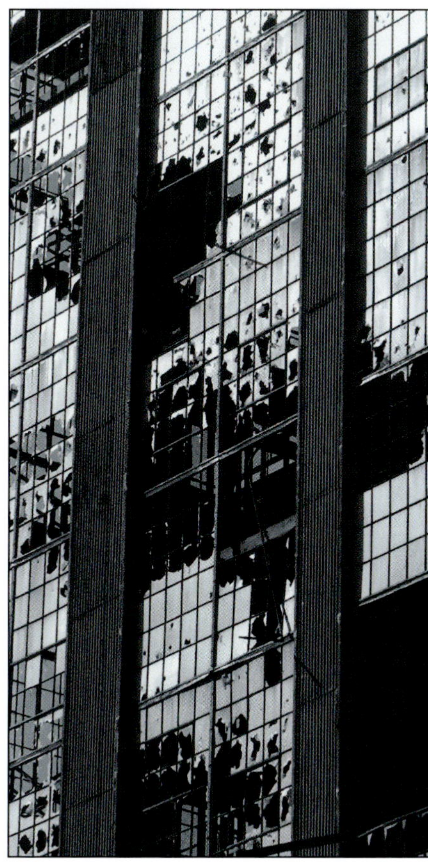

The Huber was dismantled and sold for scrap metal. The once mighty colliery was gone. Clarke has a view from his Ashley home across the street from where the breaker once stood. He says, "It was upsetting to see a lot of history going down the tube. Like the Hotel Sterling, anything of significance or value they demolish."

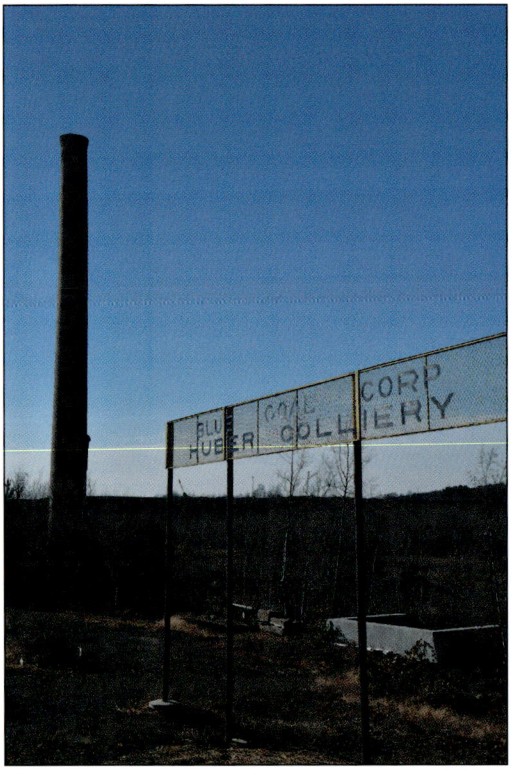

A few remnants of the colliery, like this smokestack and the Blue Coal sign, remain. In the 1990s, the Ashley Breaker Preservation Society was formed to preserve the site as a museum. The group later reformed as the Huber Breaker Preservation Society in 2001, but a lack of funding prevented members from saving the site before its demolition.

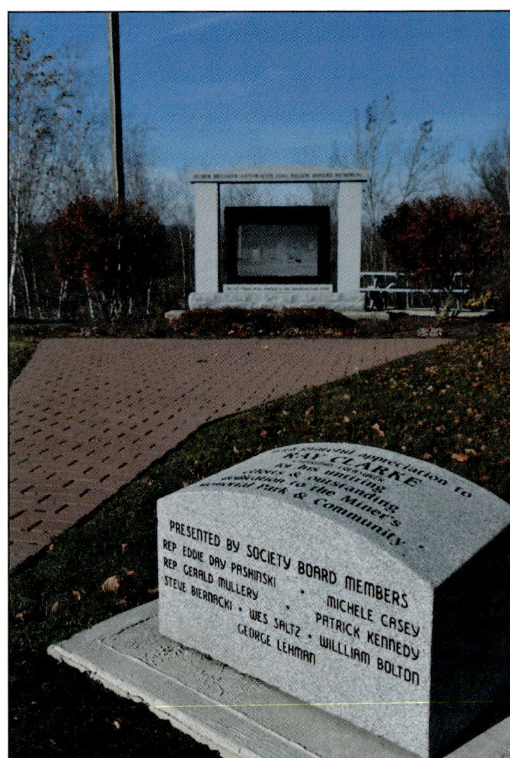

Even before the Huber fell, the nonprofit began erecting a memorial to the colliery and the workers who toiled there and in the mines. After the donation of three acres from the Earth Conservancy, the Huber Breaker Northern Coal Field Miner's Memorial Park became a reality.

Visitors to the memorial are invited to explore the site but are warned not to wander into areas that are still considered dangerous. According to Clarke, "We are putting in some walking trails and hopefully will make the parking lot bigger."

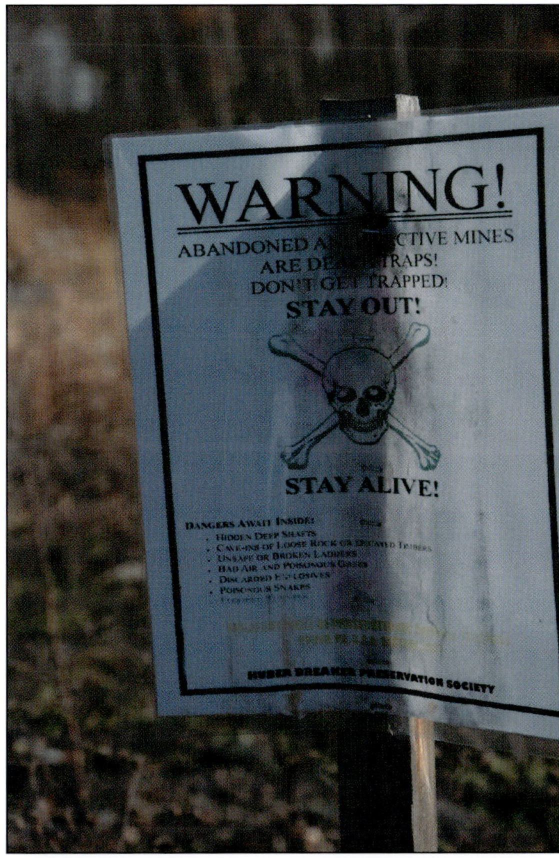

More work still needs to be done on the Huber site. According to Best, there may still be asbestos, which is considered hazardous, on the property. There is additional waste that must be removed that the Department of Environmental Protection is working to rectify.

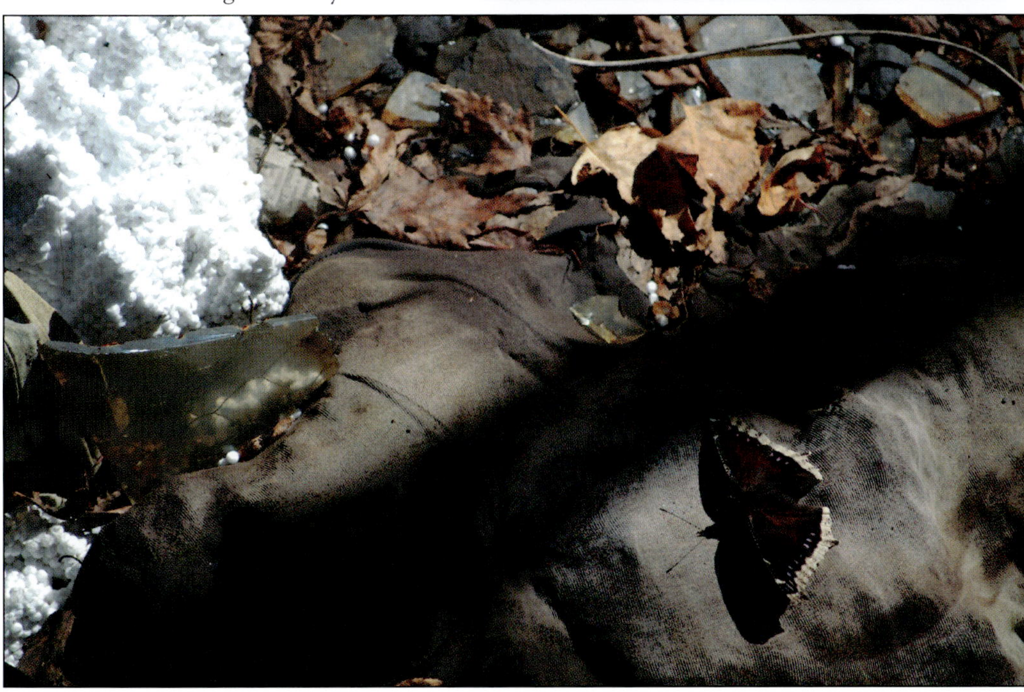

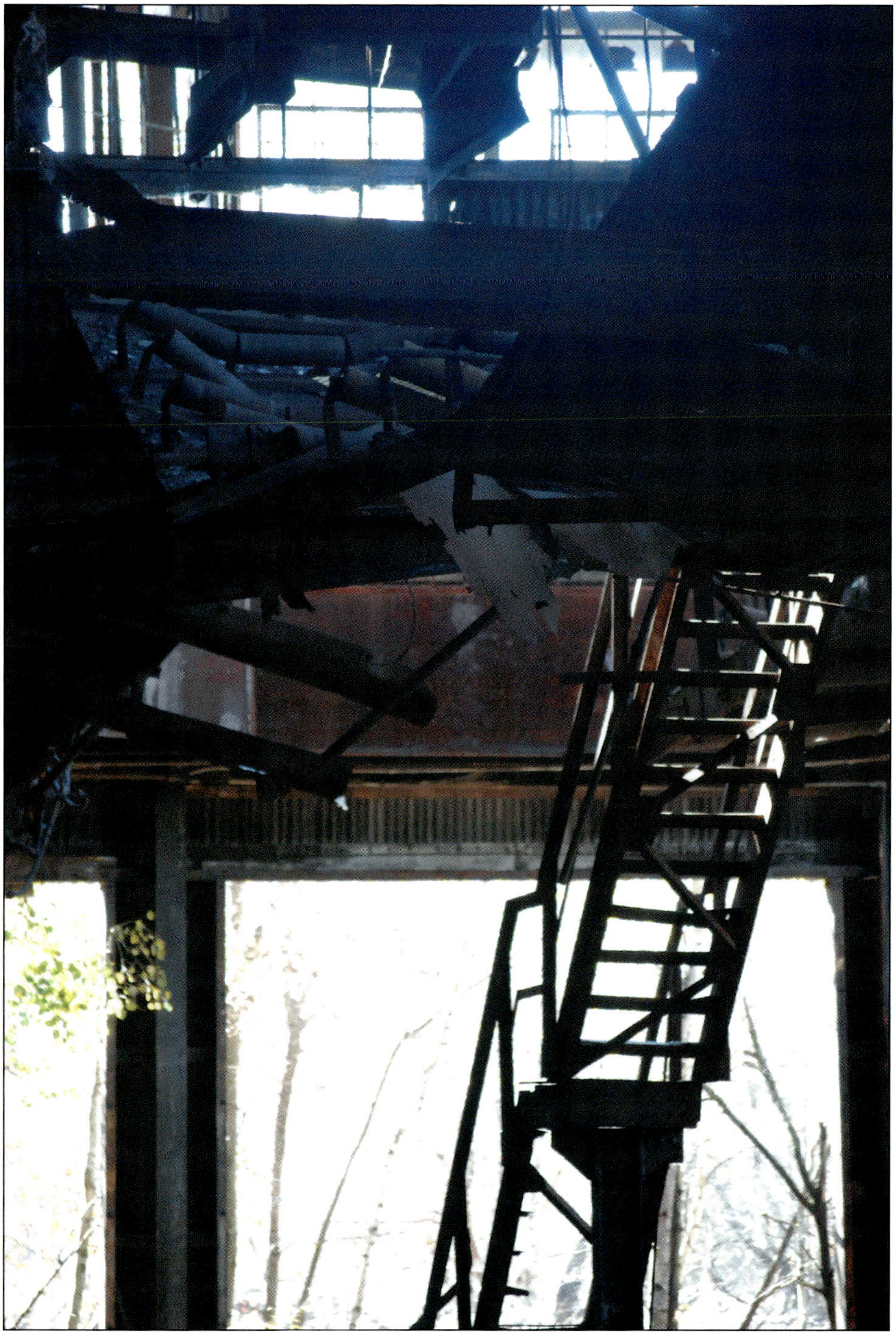

"Some people were happy to see it go because they may have lost so many people there. Some looked at it as history and others looked at it as trouble. It was controversial," says Clarke of the Huber Breaker. Though the Huber is gone, the HBPS is dedicated to preserving its legacy for future generations. Readers can visit Huberbreaker.org for more information or to donate to the nonprofit.

Three

KILLING THE CANARY

Joe Stella is a hero whose legacy should never be forgotten. He saved the lives of dozens of miners from the Knox Mine disaster despite the fact he was not even a miner himself.

It was late morning on January 22, 1959, in Port Griffith, near Pittston, when the Susquehanna River broke through a mine that lay beneath it. State law forbade mining within 35 feet of the riverbed but miners were pushed by bosses to mine closer. After mining left only 19 inches between the river and coal miners, the Susquehanna River finally gave way, and billions of gallons of water started to pour into the mine.

Thirty-five-year-old mine surveyor Joe Stella was in the mine that fated day. He had surveyed it 10 years earlier and knew to run to the Eagle air shaft. Once he got there though, he realized he needed dynamite to blast it open because it had been covered over by debris. He ran back into the mine to find dynamite and discovered a group of older miners who were struggling to escape. Best says, "Their lung capacity was limited because of miner's asthma." Younger miners had left the men behind. Stella would not make that same mistake. He saved the lives of 33 men that day. "Every minute he stayed back, he put his life in danger. That is the definition of a hero," says Best.

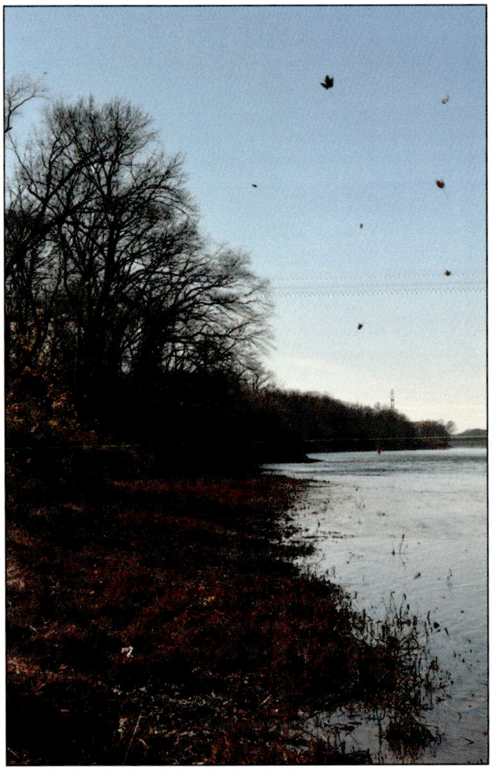

Supey was seven years old when the Knox Mine disaster happened. He says, "We could almost see the mine from our school. The principal called and said, 'Send all the children home.' Everyone's dad worked in the mines, so all the children were worried. I ran home and dad was there." Pictured here is the area where the Susquehanna breached its riverbed.

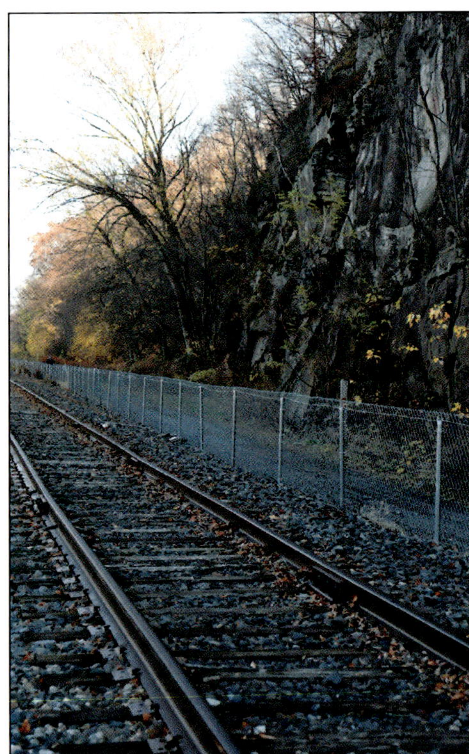

After Stella's heroic efforts, there were still men trapped in the mine. The nearby railroad tracks were rerouted to the river so railroad cars could be dumped into the whirlpool to plug the hole. The attempt was futile. "I saw them pushing gondolas into the hole and they were going down like toothpicks," says Clarke. The harrowing sight was captured on video and can be seen on YouTube by typing in "Knox Mine disaster footage."

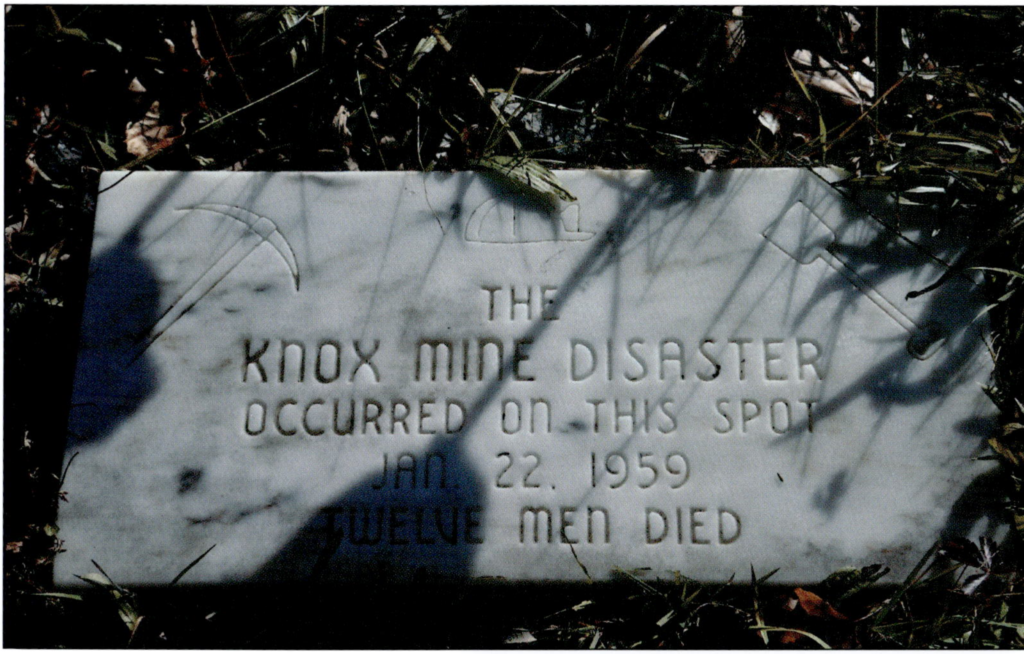

The Knox Mine disaster claimed the lives of 12 men. According to the *Times Leader*, they were Samuel Altieri, John Baloga, Benjamin Boyer, Francis Burns, Charles Featherman, Joseph Gizenski, Dominick Kaveliskie, Frank Orlowski, Eugene Ostroski, William Sinclair, Daniel Stefanides, and Herman Zelonis. Their bodies were never recovered. No executives of the Knox Mine Coal Company were imprisoned for their deaths—or what some call their murders.

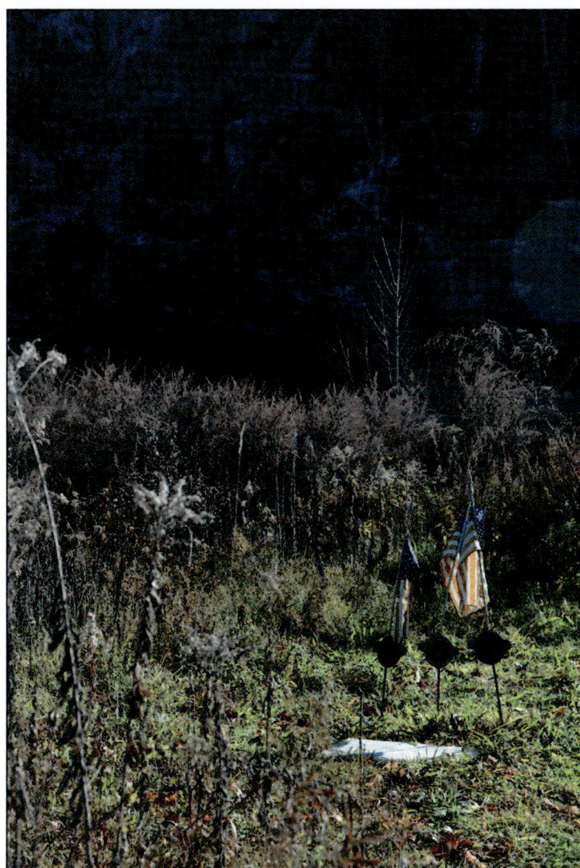

A memorial to the miners marks the spot of the disaster that ended underground coal mining in the Wyoming Valley. "Nobody wants to admit it, but it was a good excuse for the coal companies to go out of business because they were large coal companies and they were subletting all their coal out to subcontractors. The companies were paid royally by the government because the government couldn't get in to see what they lost," says Supey.

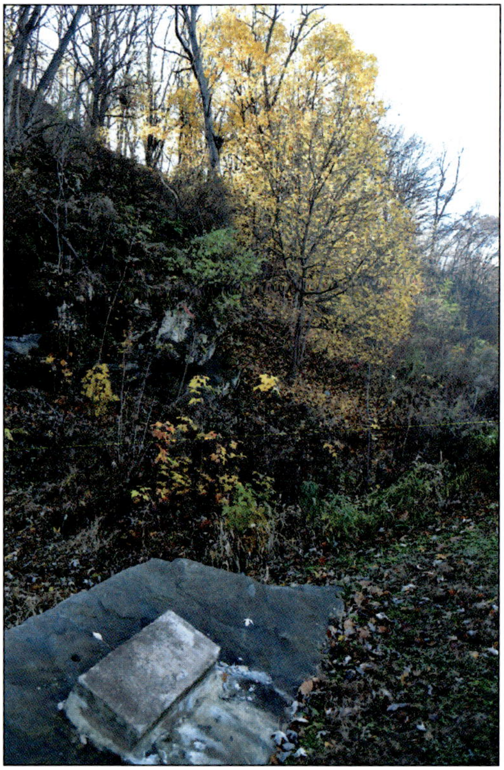

Along the railroad tracks, a second memorial was placed marking where Stella, along with others, led miners out of the Eagle air shaft. Sadly, nothing of it remains except its outline where it was once attached to a boulder. Amadeo Pancotti received the Carnegie Medal for aiding in the rescue of 33 men from the shaft. Unfortunately, Stella never received one.

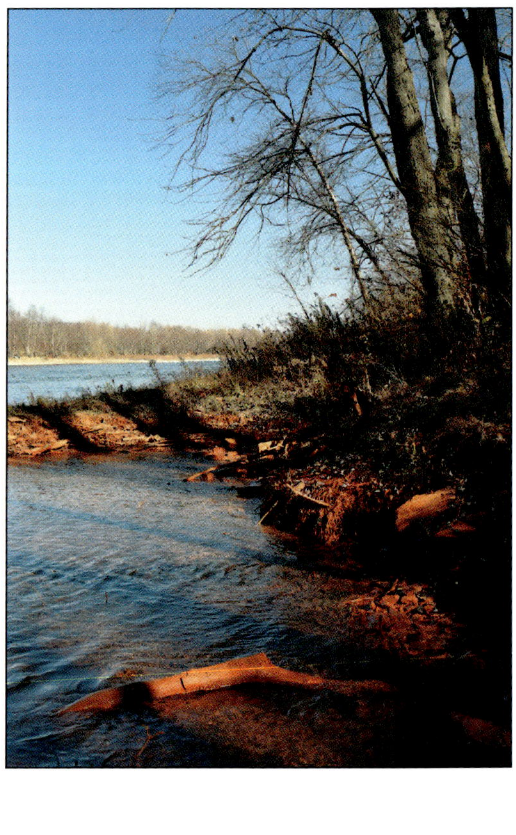

The banks of the Susquehanna River seemingly still bleed for those 12 miners nearly 60 years later. Located near the memorial, this shoreline spews rubiginous water from the mines. This is actually acid mine drainage, which is a problem affecting water quality in the region. The problem is being addressed throughout the area by the Eastern Pennsylvania Coalition for Abandoned Mine Reclamation (EPCAMR), which is a regional environmental nonprofit organization.

EPCAMR executive director Robert Hughes says, "EPCAMR has been trying to secure funding from state, federal, and private foundation sources to both monitor the chemistry, determine the extent of the abandoned mine drainage, and to map the complex underground mine pool system that lies beneath the Wyoming Valley."

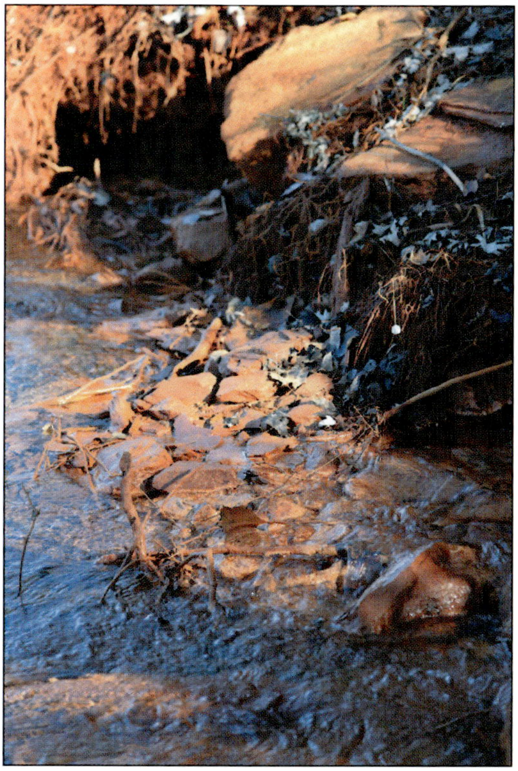

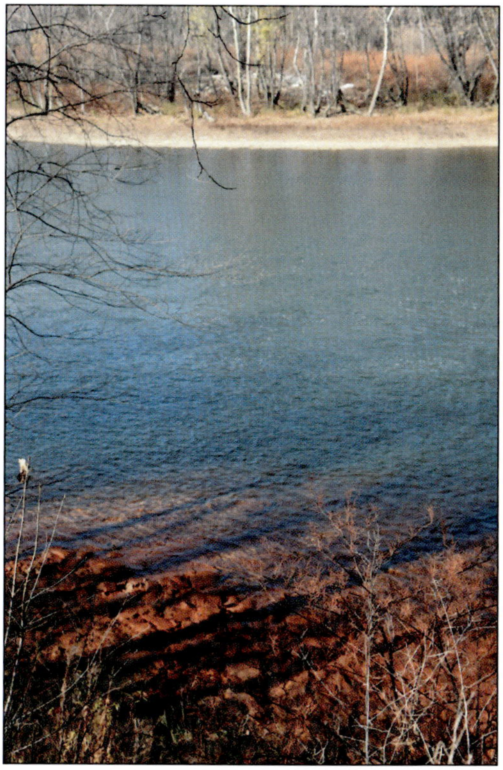

Of all the sites in this book, the Knox Mine disaster area is perhaps the hardest to locate. The best point of reference is to locate Eli's Sales and Service at 1425 Main Street in Jenkins Township. Across from the business is a trailer on the corner of a road that will lead to public parking. Visitors can head to the Luzerne County Rail Trail and walk north for 10 minutes. The memorial will be across the railroad tracks beside the Susquehanna River.

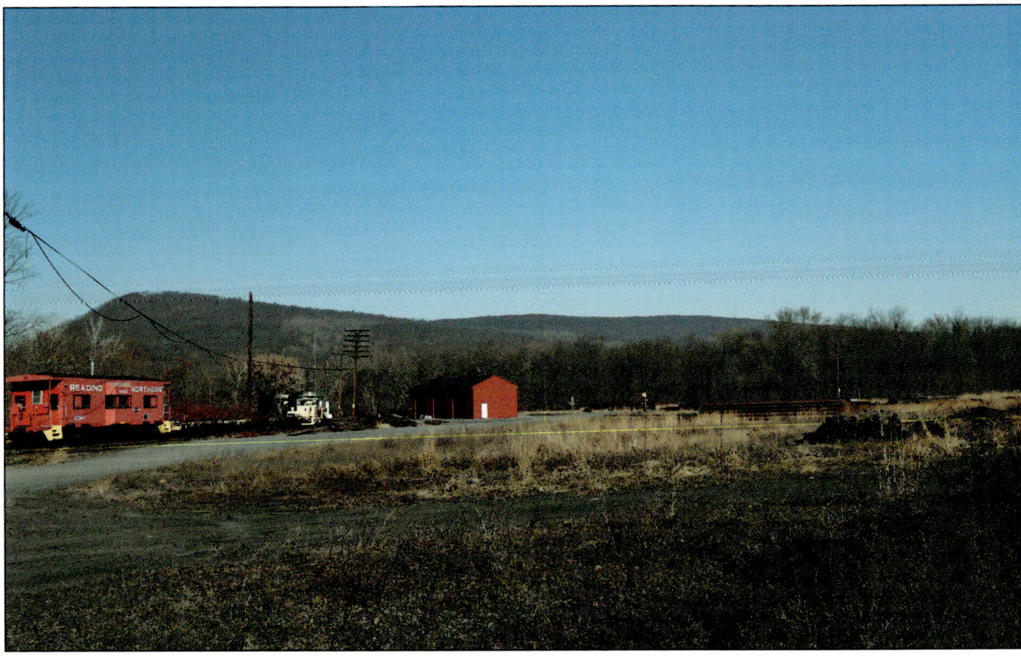

Not far from the Knox Mine disaster is the site of the Twin Shaft disaster, which was one of the worst mining accidents in Pennsylvania. On June 28, 1896, in Pittston, 58 boys and men lost their lives below the Newton Coal Company's Twin Shaft Colliery.

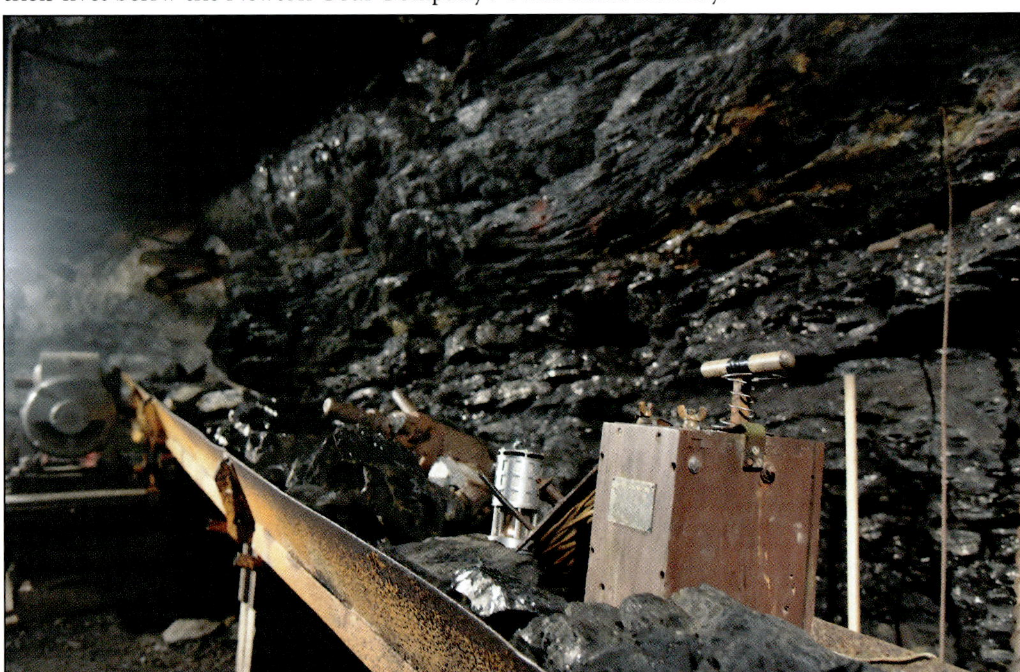

"There were a thousand ways to die in a coal mine," says Supey. The leading cause of injuries in the mines was roof falls. They occurred when props failed or explosives triggered bigger collapses. Additional hazards included losing fingers, arms, and legs along with electrocution, drowning, and burning.

"The only way they learned about accidents was from seeing them. If you were a farmer in Czechoslovakia, you came over here and when you saw someone get their arm or leg cut off, you learned that I am not going to let that happen to me. Men were getting killed left and right because they didn't put supports up. There were no mining schools, only the school of hard knocks," says Supey.

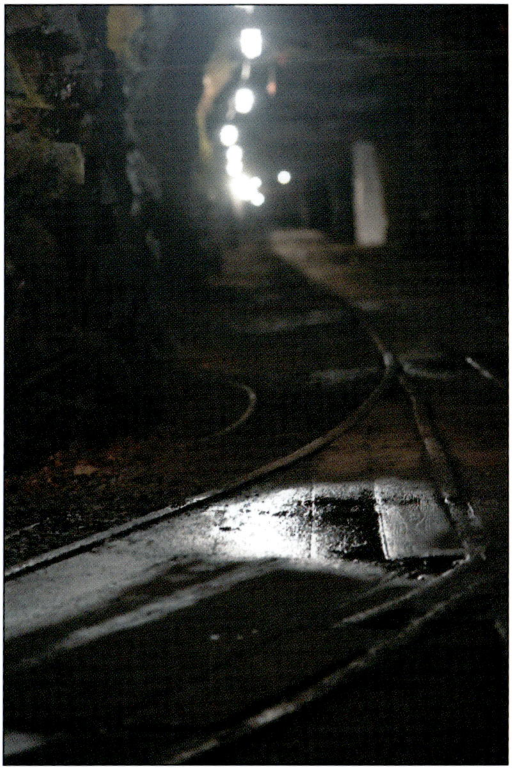

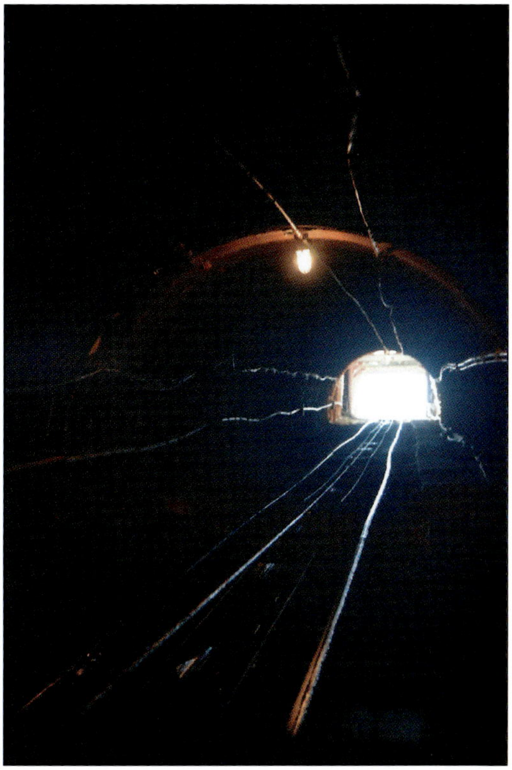

Working underground for the miners, laborers, and children meant traveling into the coal mine before the sun arose and, often, leaving long after it had set. Daylight, or any light for that matter, was a scarcity in their lives. While working in the mines, there was no electricity, so workers had only headlamps to light their way.

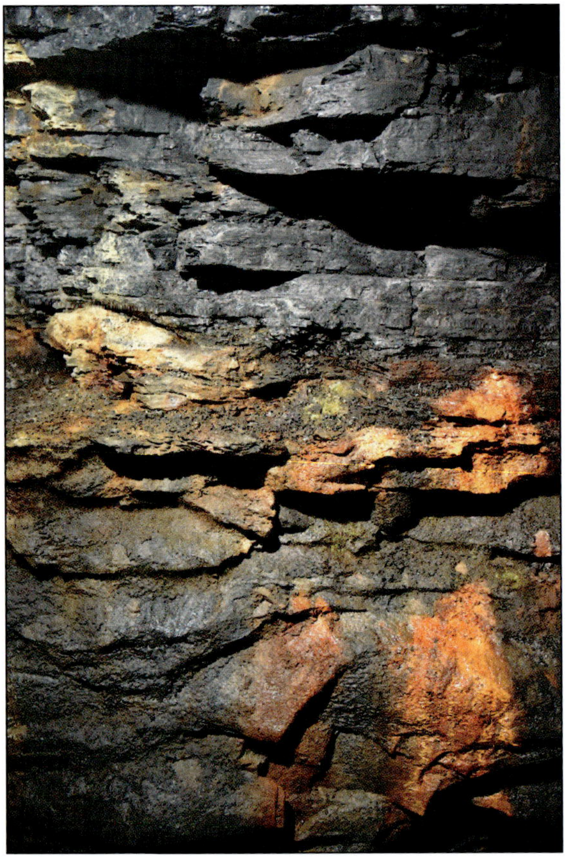

Dangerous air quality conditions were another hazard of mining. Miners were often accompanied by canaries, which were more sensitive to harsh gases. The canaries chirped their days away in the mines, but when the chirping ceased, it was a low-tech warning to miners that it was time to run. Canaries are sensitive to toxic gases because of their respiratory systems. For every one breath a human takes, the canary takes two.

Mice and rats were also good indicators of toxic gases. If a hoard of rats or mice were seen fleeing a shaft, it was another sign that impending doom was around the corner. Both rodents were also adept at stealing miners' lunches. Pictured here is a reproduction of a miner's lunch box, which is sealed tightly to protect its much-needed contents to get a worker through a long day.

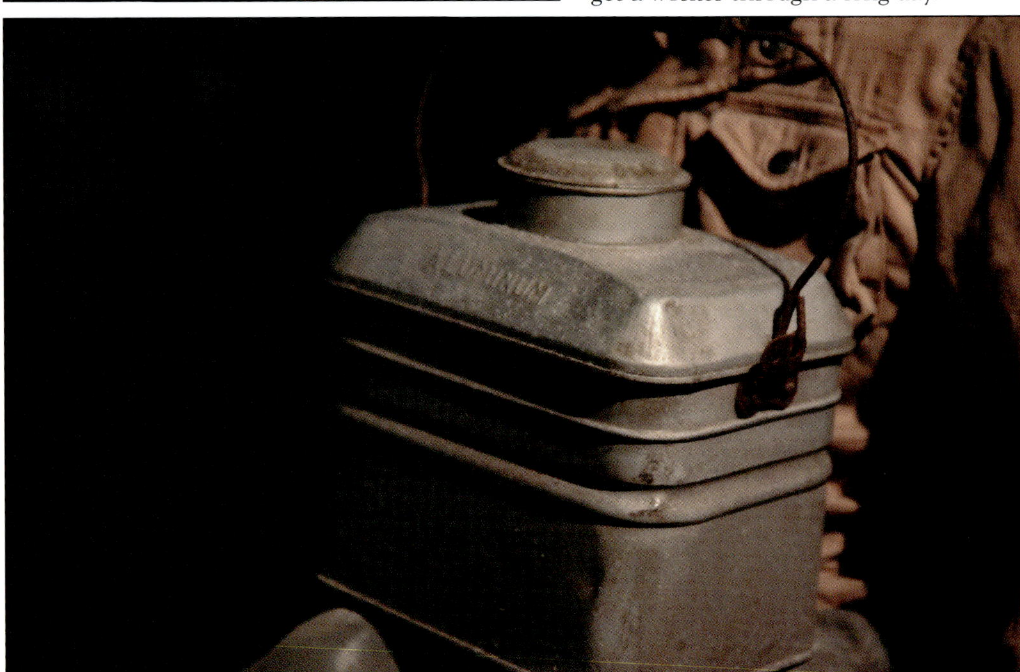

Keeping air quality at its optimum was the job of fans like this one pictured at the Lackawanna Coal Mine. Located aboveground, this fan is capable of pulling 22,000 cubic feet of air through the mine. The fan's task was to suck air out of the mine shaft with fresh air being pulled from the shaft's entrance.

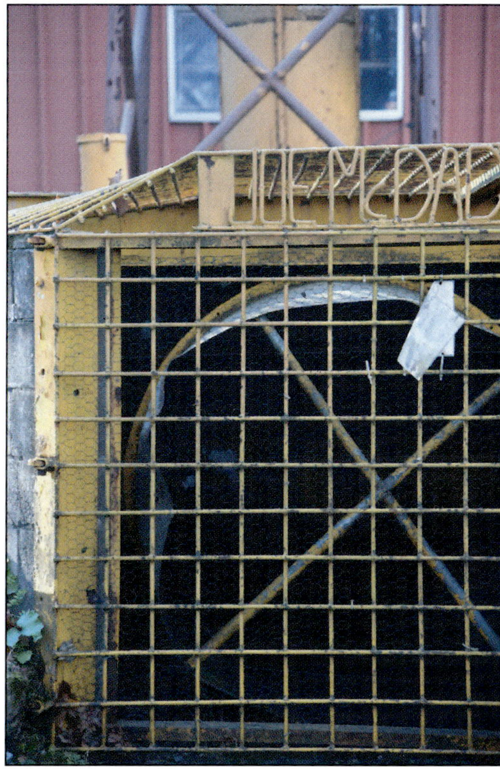

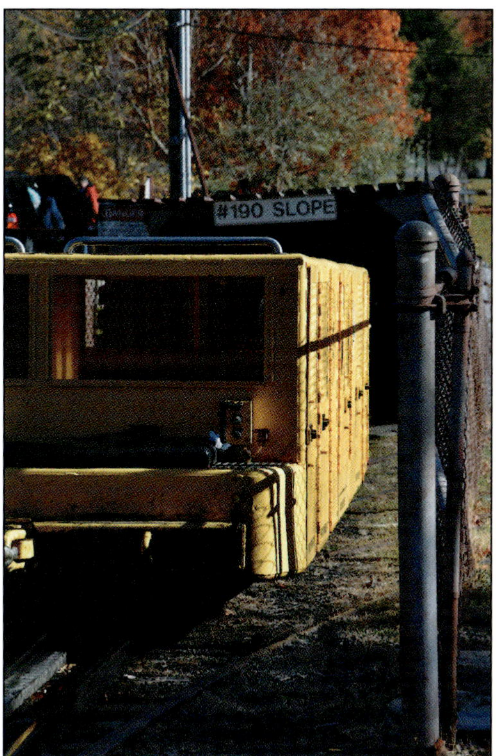

When Supey started restoration on Slope No. 190 at the Lackawanna Coal Mine in Scranton, air quality was the first issue to be rectified. Supey says, "It was oxygen deficient. Instead of it being 16 percent oxygen it was only 5 percent or 8 percent air; 12 percent is legal. You can't work below that. It's called black damp when air has an insufficient amount of oxygen in it."

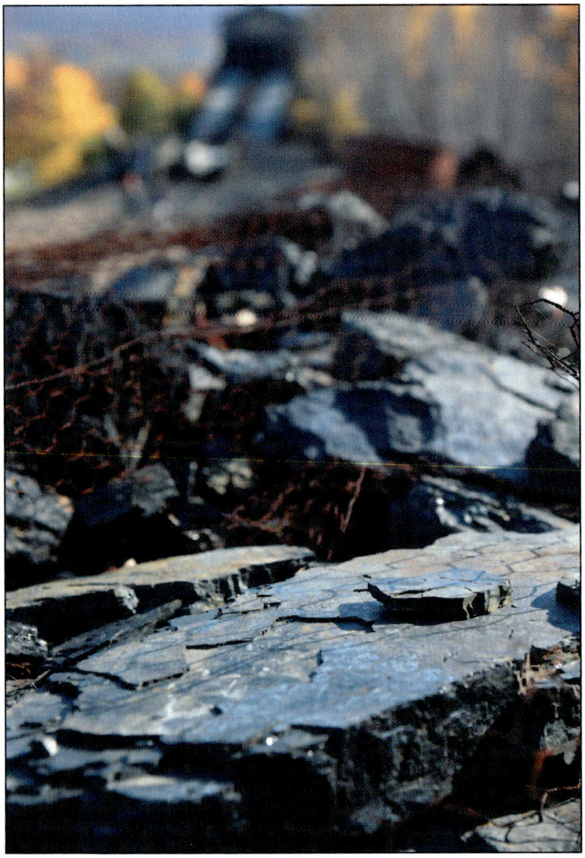

Anthracite coal can be found around the world, but it is only found in such abundance in Northeastern Pennsylvania. Anthracite is superior to bituminous coal because of its ability to burn at higher temperatures due to its high carbon content. Despite the tons of anthracite that have already been mined, there are still millions of tons of unmined anthracite sitting beneath the ground.

The coal mining industry in the 1800s and early 1900s lured men from around the world to move to Northeastern Pennsylvania. "If you came [here], you either worked in the mines, the breaker, or in a repair shop that the mines owned," says Supey.

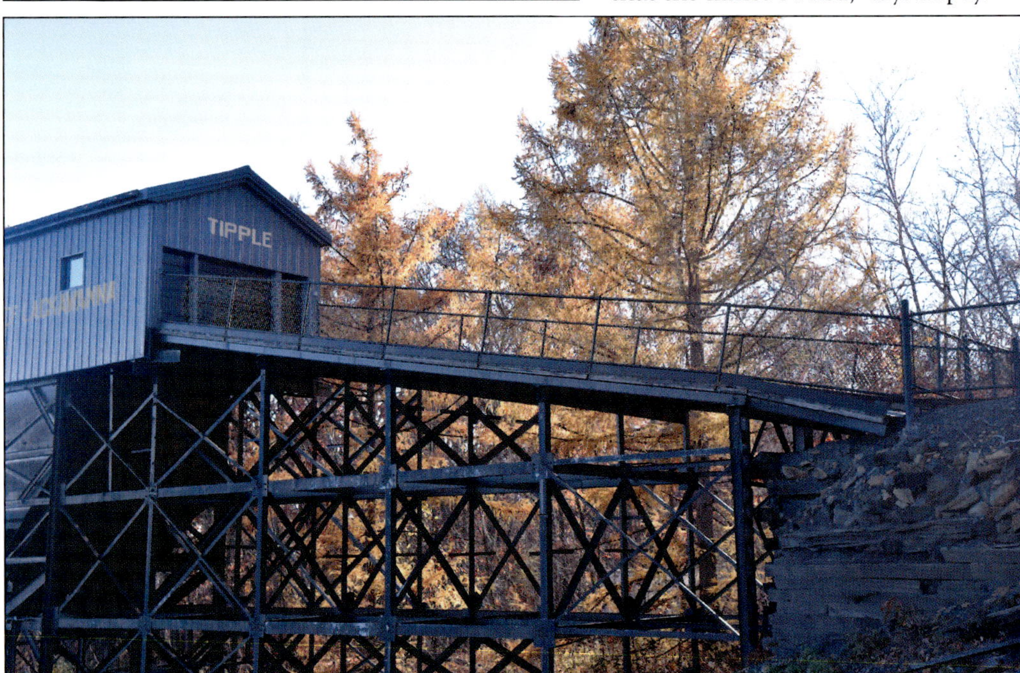

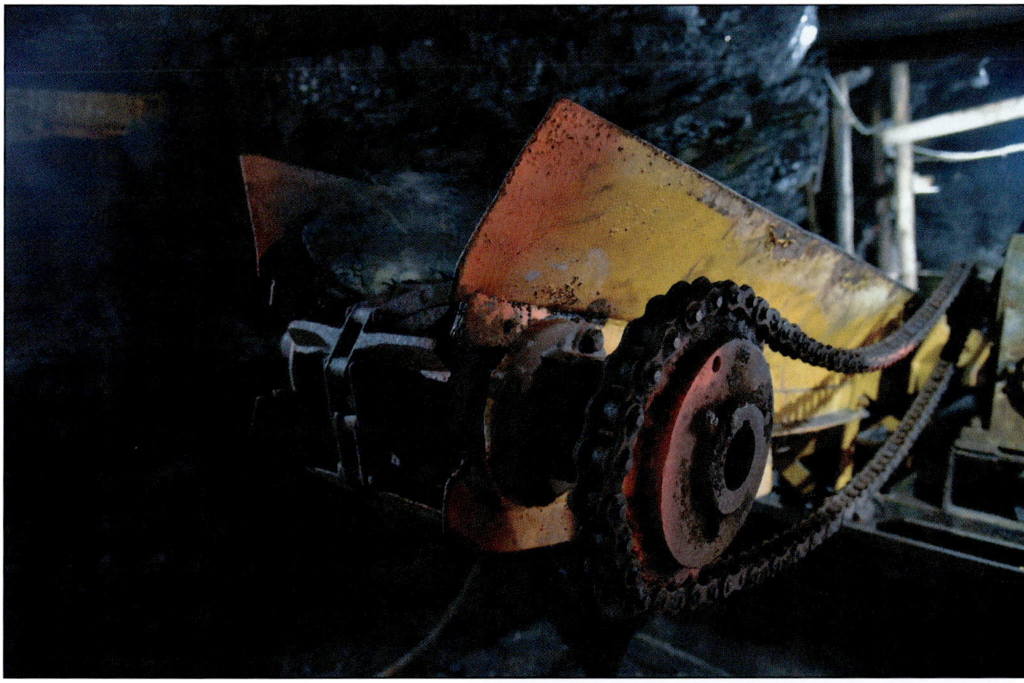

The top-paying job underground was miner, but the position did not always go to the most-skilled applicant. "Like anything, it was corrupt. You could go to Joe's bar here on the corner and get mining papers if you knew somebody. To get a job in these mines, you had to know somebody or give someone a kickback. Being related helped too," says Supey.

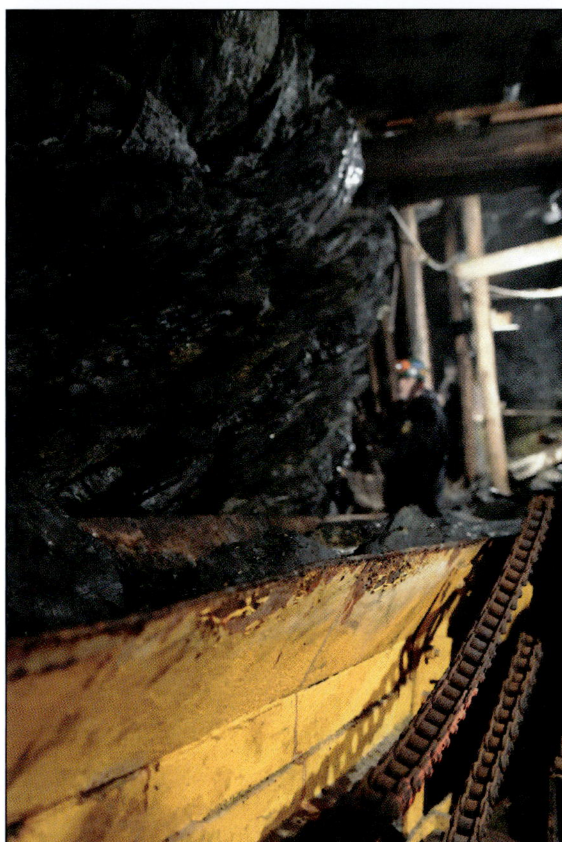

No matter what job a person performed in the anthracite mines, the pay was a pittance. Supey says, "It was almost slavery. They had to buy everything from the company store and they had to live in a company home. When they saw you doing a little better, then they would raise everything. They had them in a grip and they could never get ahead."

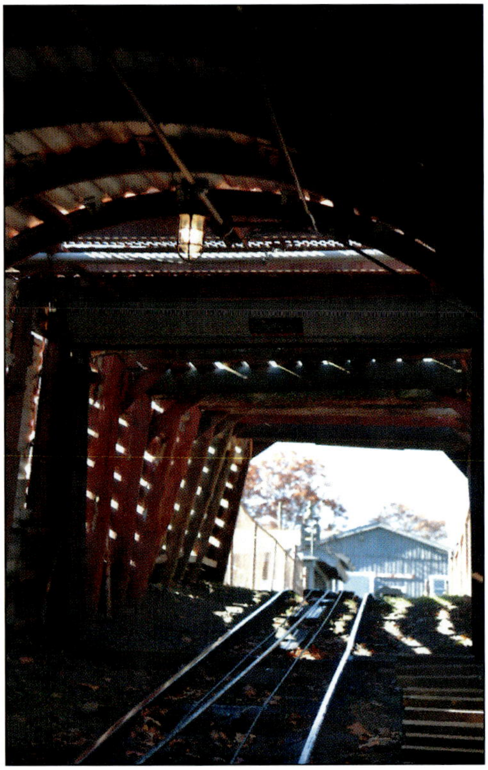

A visit to the Lackawanna Coal Mine Tour drops visitors 300 feet belowground where they embark on a guided tour by former miners. Slope No. 190 was formerly owned by the Continental Coal Company, which operated the mine for over a century until its closure in 1966.

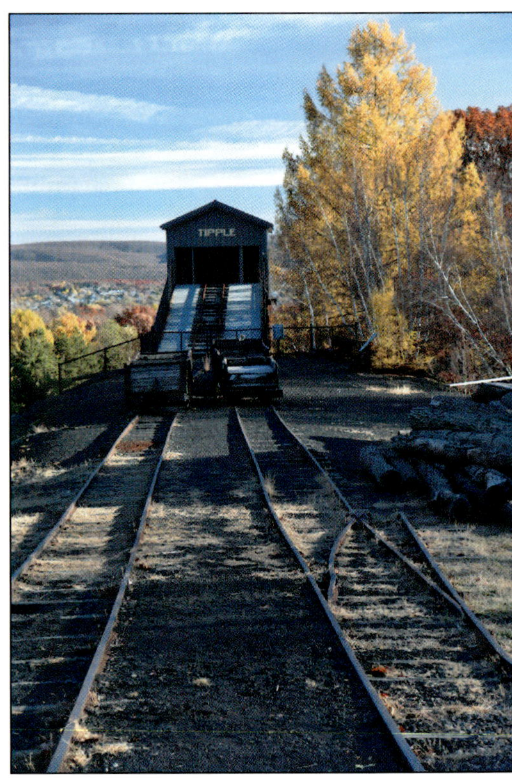

One of the notable sights at the Lackawanna Coal Mine Tour is this tipple. The tipple received loaded cars of coal and rock mixture, which was then tipped over at the top of the structure and dumped into a truck or train below. It was then transported to the Taylor Breaker where it would be cleaned, crushed, and sorted into various sizes.

One of the few benefits of working in a coal mine was the year-round temperature of an average of 50 degrees. This made the mine shaft a better place for workers to repair coal cars. Pictured here is the motor pit with a chain hoist where repairs were easily handled away from the cold or heat of the outside.

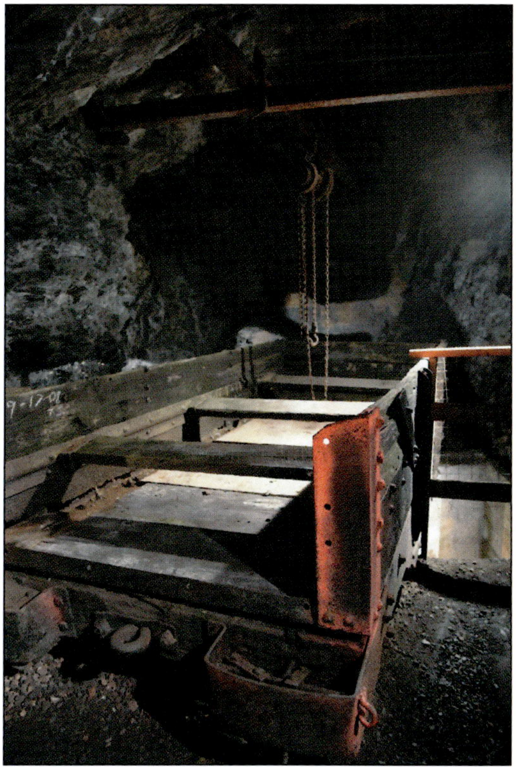

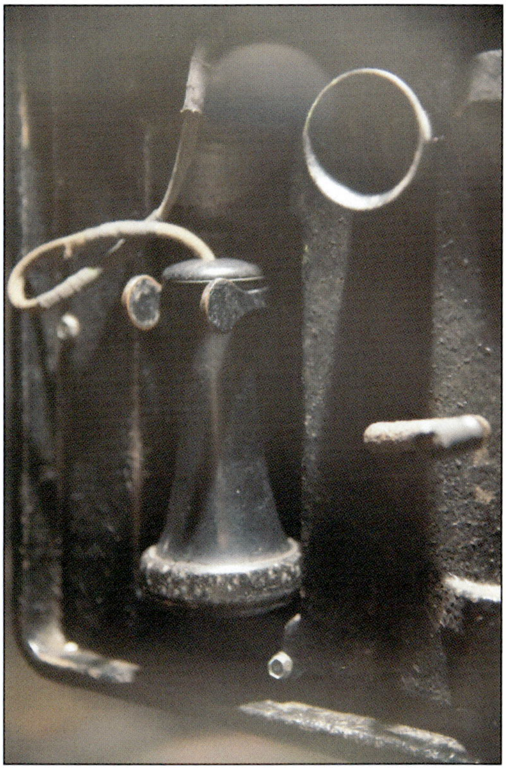

Phones were used only in larger mines. The device was powered by batteries and used for communication within the mine and with the outside hoist operator. The one pictured here in Slope No. 190 is a Bell Telephone from around the 1930s and was placed here during renovations.

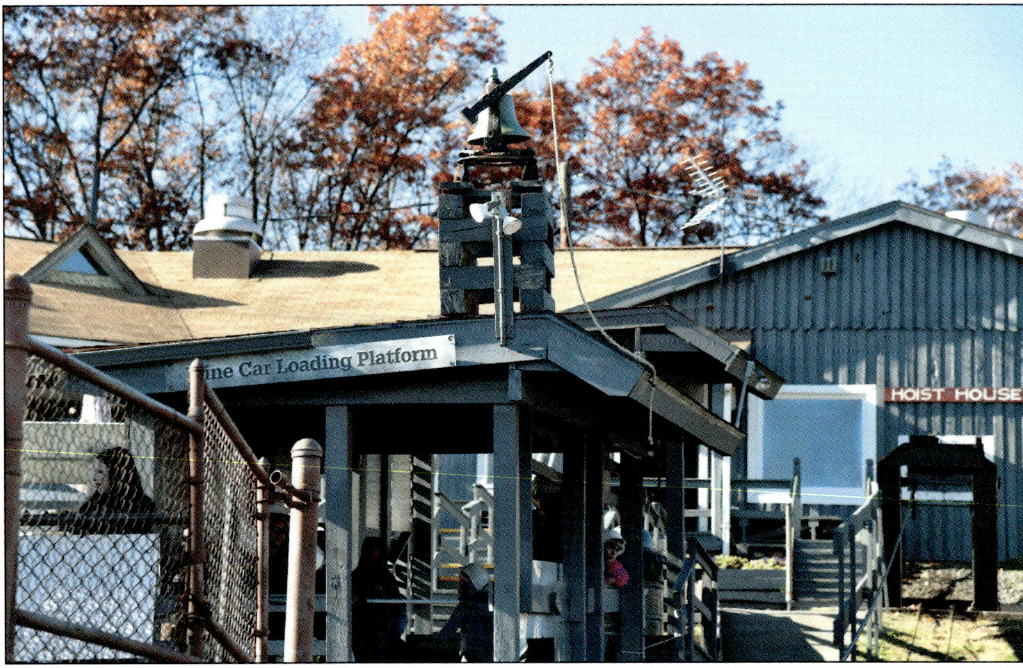

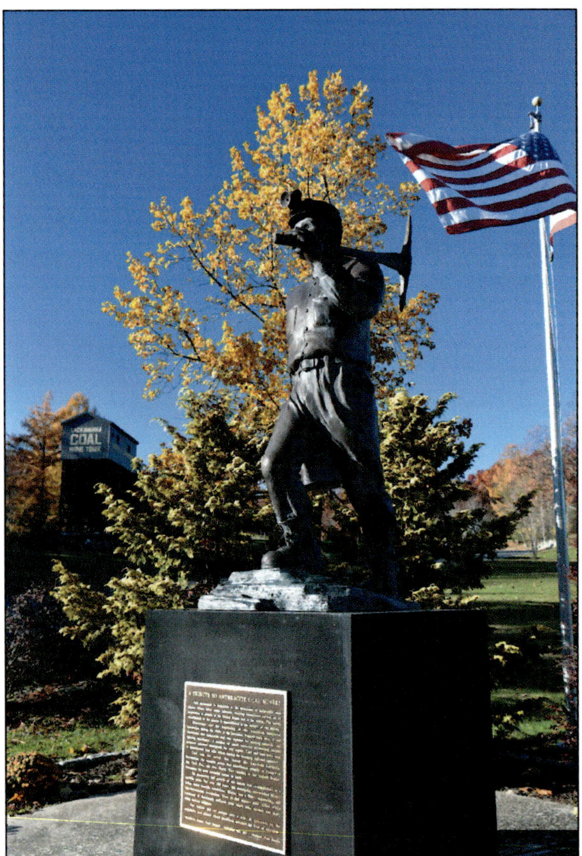

Supey and his father, Tom Supey Sr., were responsible for restoring Slope No. 190 over five years. Once it was completed, the younger Supey served as a tour guide for 30 years. He says, "We used to have 35,000 school kids come through the tour each year. They came from New York, Delaware, Maryland, and New Jersey and left with an understanding of a life that doesn't exist anymore."

This memorial to miners at the Lackawanna Coal Mine pays tribute to the thousands of miners who lost their lives in mine accidents or to black lung disease. The memorial that stands at the entrance to the coal mine was sculpted by Frank Talarico. In 2016, for the first time, Pennsylvania designated the third Saturday in July as Coal Miner Heritage Day.

Four

Rise of the Rails

Without the codependent relationship of steam locomotives and anthracite coal, the Industrial Revolution would have been miserably crippled. It is hard to overstate how both industries fueled the propulsion of the United States' development in the 1800s.

"Railroads, the first major industry in the United States, made possible the growth of industries like coal, steel, flour mills, and commercial farming. They established cities like Chicago and had an impact on urban design," says the Gale Group, which is an online educational resource.

Scranton was one of those cities that flourished once a railroad was established. Prior to the rails' arrival, the local economy was an agrarian one. Railroads changed that dynamic, allowing Scranton to grow into one of the powerhouse cities of the Industrial Revolution.

Steam railroads shortened distances between cities, allowing for the region's anthracite to be carried to other growing metropolises like New York City and Philadelphia. The transportation of coal is what ultimately made it the dominant fuel source, replacing water and wood power.

Today, it is hard to grasp the impact steam railroads and coal had on America since today's interaction with both is limited. However, the Steamtown National Historic Site in Scranton does an exceptional job of educating the public on the vital role both played in past lives.

Not to diminish the power of words, but nothing can replace touching the steel of a steam locomotive or stepping into a caboose from the 1800s. Gratefully, this is one piece of history that has yet to be lost, thanks to the National Parks Service.

The bulk of the photographs in this chapter were taken at Steamtown.

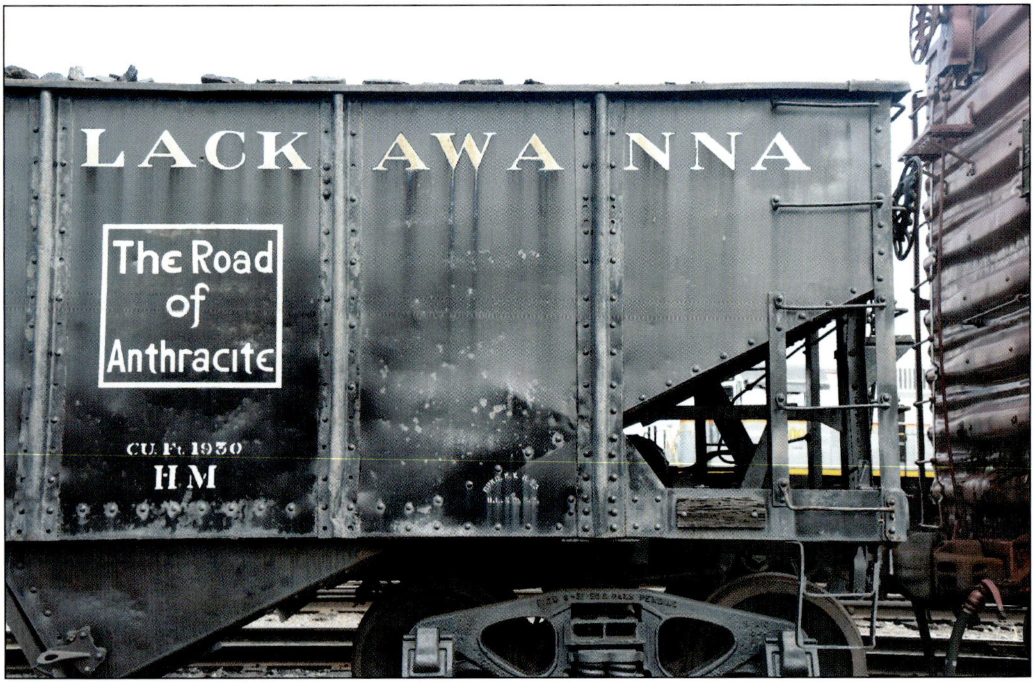

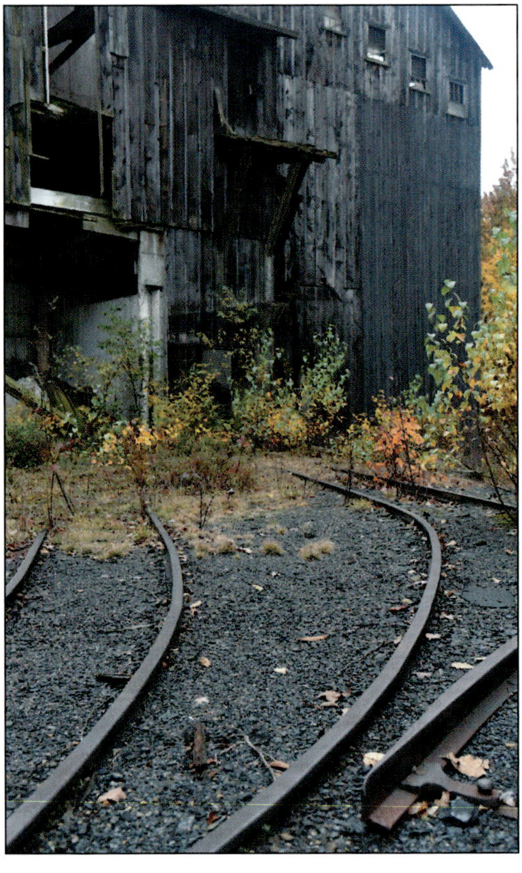

The road to anthracite was built upon steel rails. The arrival of the Scranton family from Connecticut launched Scranton into the Industrial Revolution. In 1840, geologist William Henry surveyed Northeastern Pennsylvania for a suitable location to produce iron. He ultimately settled on a 500-acre plot that would later become the city of Scranton.

Henry turned to his son-in-law George Scranton and George's brother Selden and the Scrantons' cousin Joseph Scranton for financing. After capital was acquired, the family constructed a blast furnace and began mining ore and coal to make iron. Despite almost facing bankruptcy, the family went on to become the first Americans to mass-produce iron T-rails after establishing the Lackawanna Coal and Iron Company.

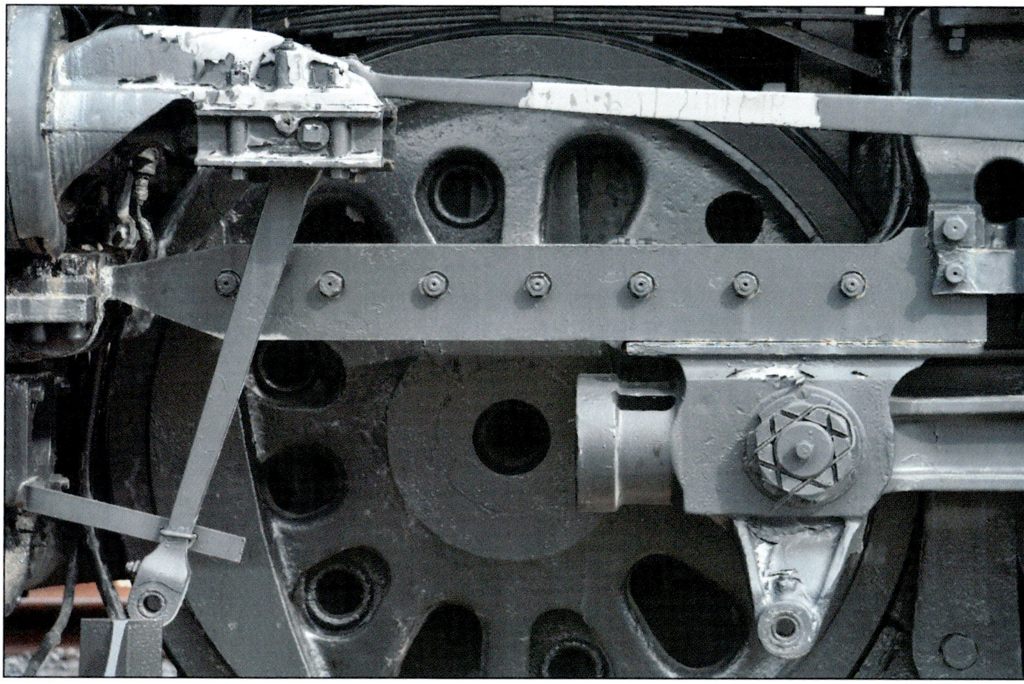

Producing rails for other railroads led the Scrantons to establish their own railroad. The Delaware, Lackawanna & Western Railroad (DL&W) was created in 1853 to transport iron products and anthracite coal. In addition to transporting coal, the railroad also owned coal mines.

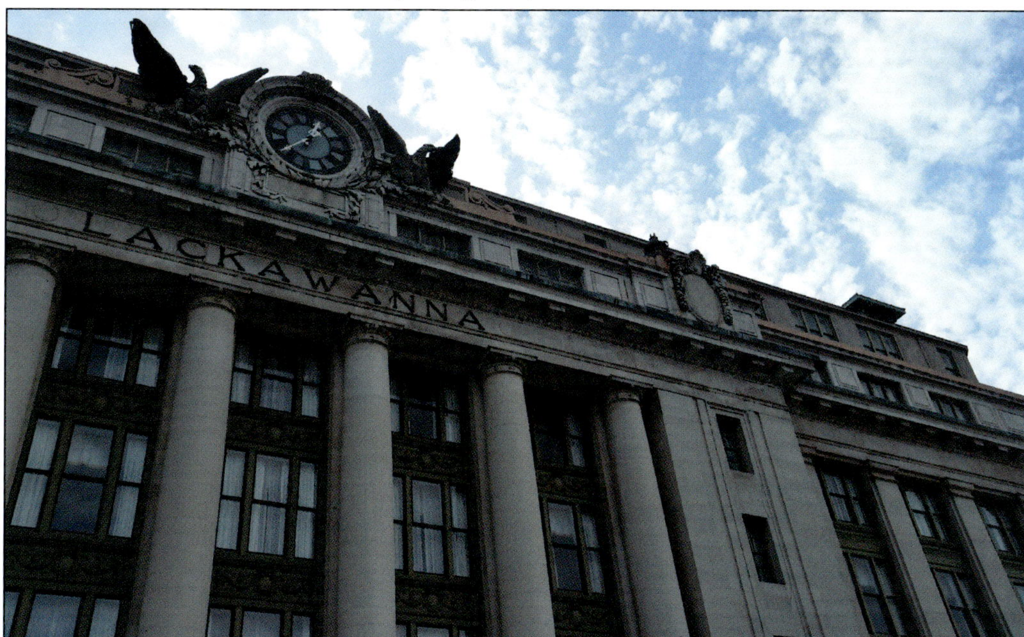

Offices for the DL&W were housed in the Lackawanna Station in Scranton, which cost $600,000 to build in 1908 and was designed by New York architect Kenneth Murchison. The building's main function was as a train station. Its last passenger train departed in 1970, leaving the structure vacant for over a decade.

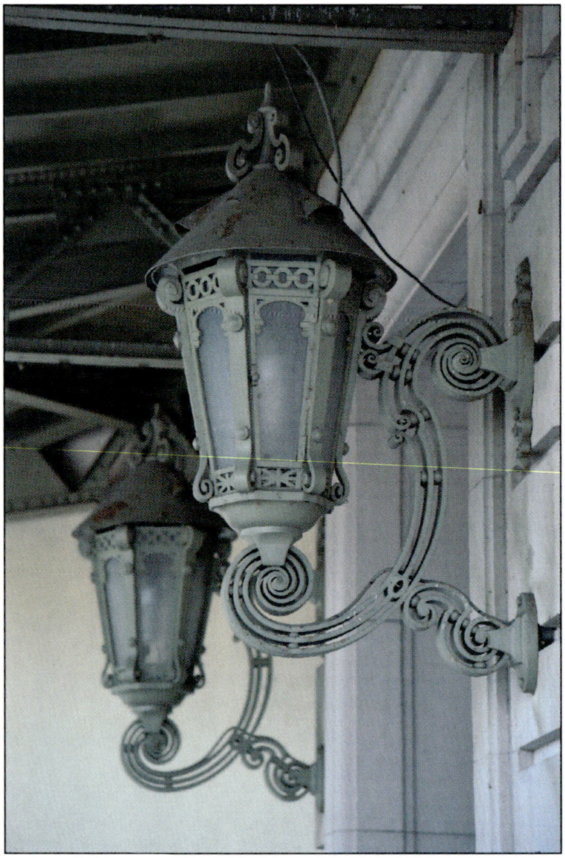

The Classical Revival structure got a second chance to shine in 1983 after a $13 million renovation that transformed it into a hotel. Today, it is one of Scranton's architectural jewels. Visitors to the hotel can get a glimpse of a bygone architectural era while enjoying a Sunday brunch indoors or cocktails on the patio.

Scranton grew tremendously as a result of its booming economy. In 1851, Scranton officially became a city and grew to be known as the Electric City. The name celebrates the city's distinction of having the first electric-operated trolley system in America, which was developed in 1886. While the trolley system no longer exists, the Electric City Trolley Museum has preserved its history.

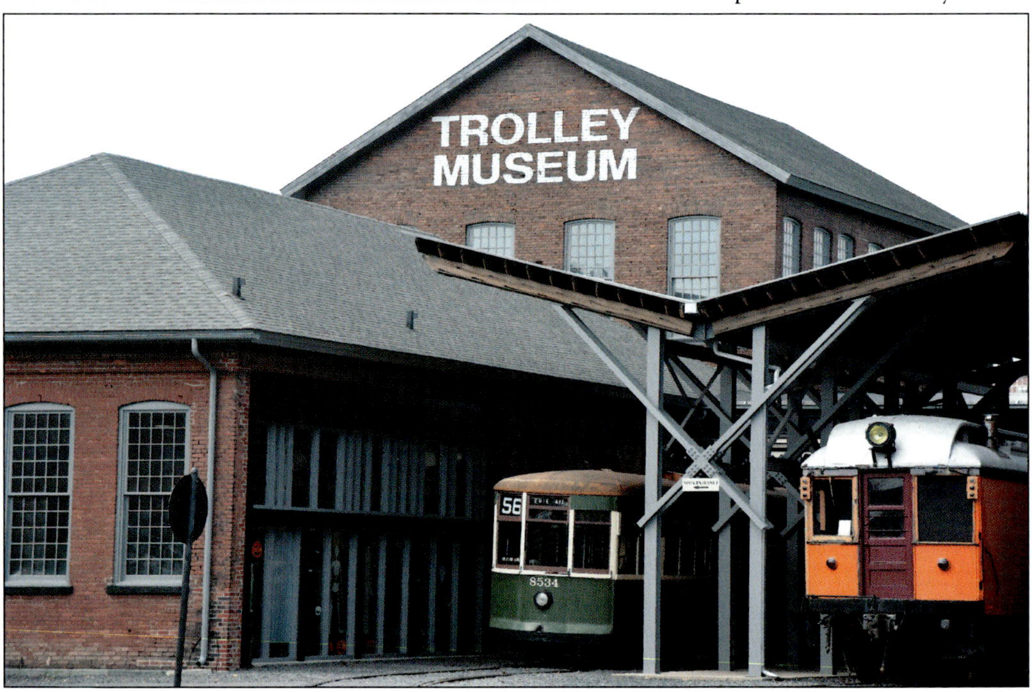

The city and its growing industries attracted workers from around the world. According to *Wikipedia*, "From 1860 to 1900, the city's population increased more than tenfold. New immigrants included Irish, Italians, Polish and Germans."

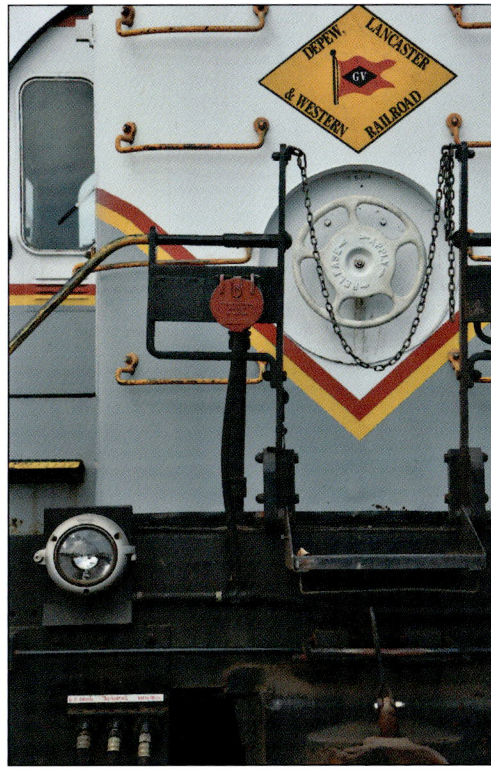

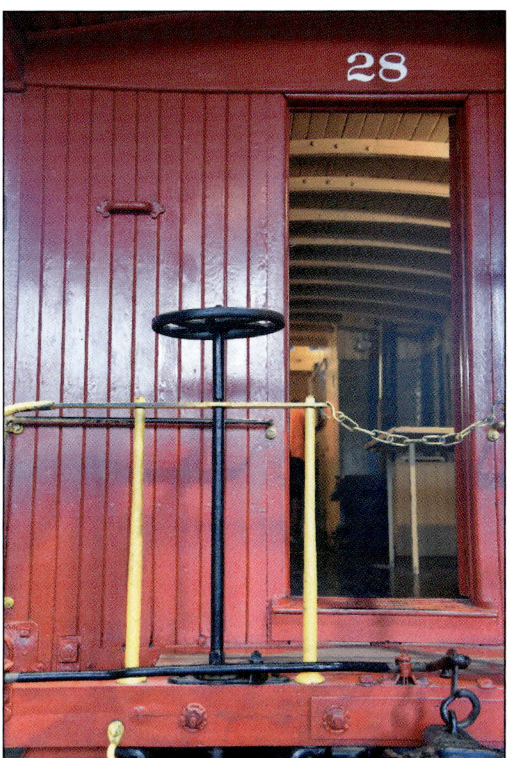

The railroad industry expanded quickly throughout the nation. In 1880, the United States had 94,000 miles of track. By 1918, that number jumped to 254,000 miles, with 65,000 steam locomotives traveling the rails.

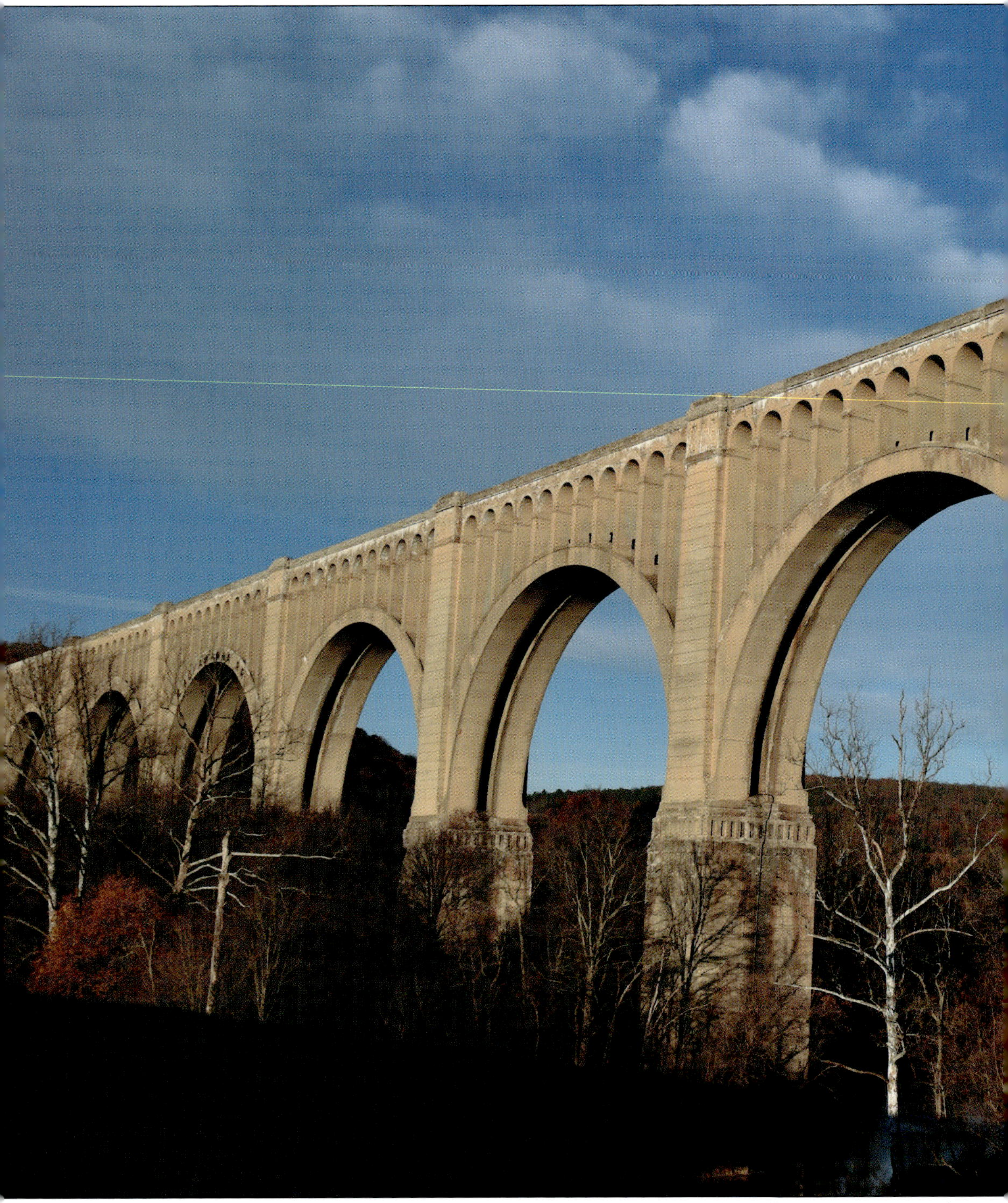

Engineering marvel Tunkhannock Creek Viaduct is over 100 years old and is still in use. Constructed by the DL&W in Nicholson, it is also known as the Nicholson Bridge or Tunkhannock Viaduct. It is 240 feet high and 2,375 feet long and crosses the Tunkhannock Creek. Construction started on the span in 1913, with 500 workers toiling 24 hours a day to complete it by 1915. It was a

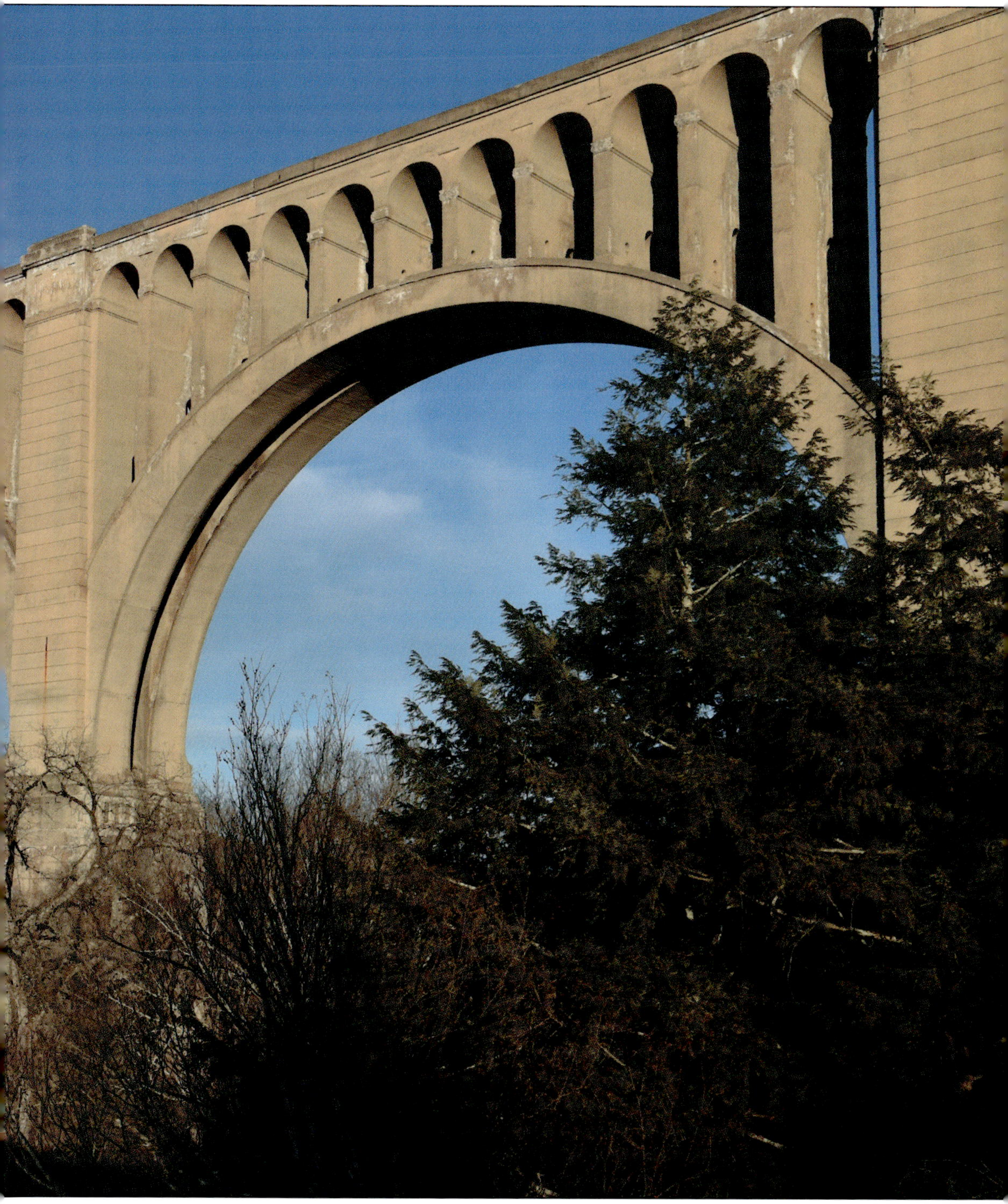

daunting task since workers had crude construction equipment that included steam shovels, dynamite, and a cement mixer. At least four workers lost their lives during construction due to falls or, in one instance, because of a dynamite explosion. The bridge is listed in the National Register of Historic Places.

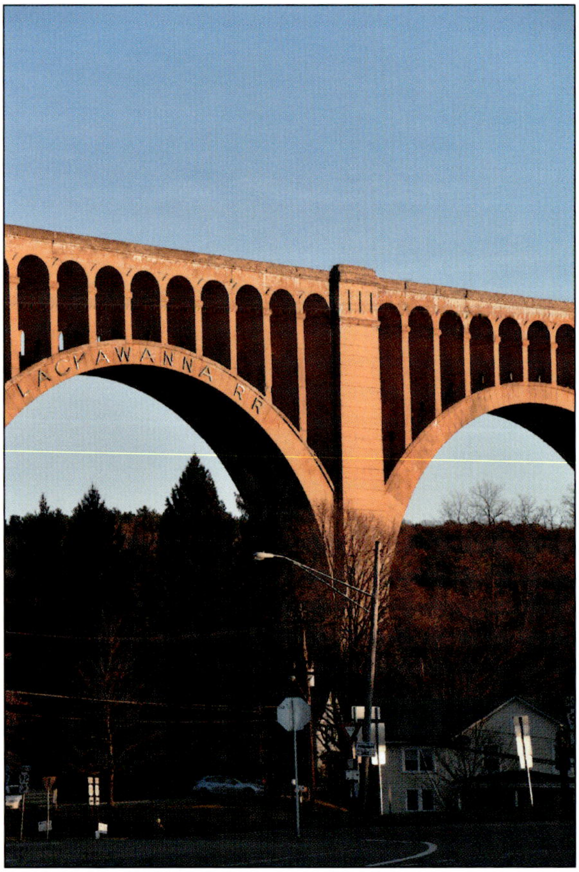

The DL&W constructed the Tunkhannock Creek Viaduct to modernize its train routes. It shortened travel by 3.6 miles between Scranton, Pennsylvania, and Binghamton, New York. The cost was $1.4 million to build the bridge but saved the company $14 million because of the shortened commute. Today, Canadian Pacific and Norfolk Southern utilize the span.

Not far from the viaduct is the DL&W Freight House. Constructed around 1849, it was the busiest station between Scranton and Binghamton. Currently, the Nicholson Heritage Foundation is planning to restore the structure and use it as a visitor's center. For information on the project or to donate, readers can visit Nicholsonheritage.org.

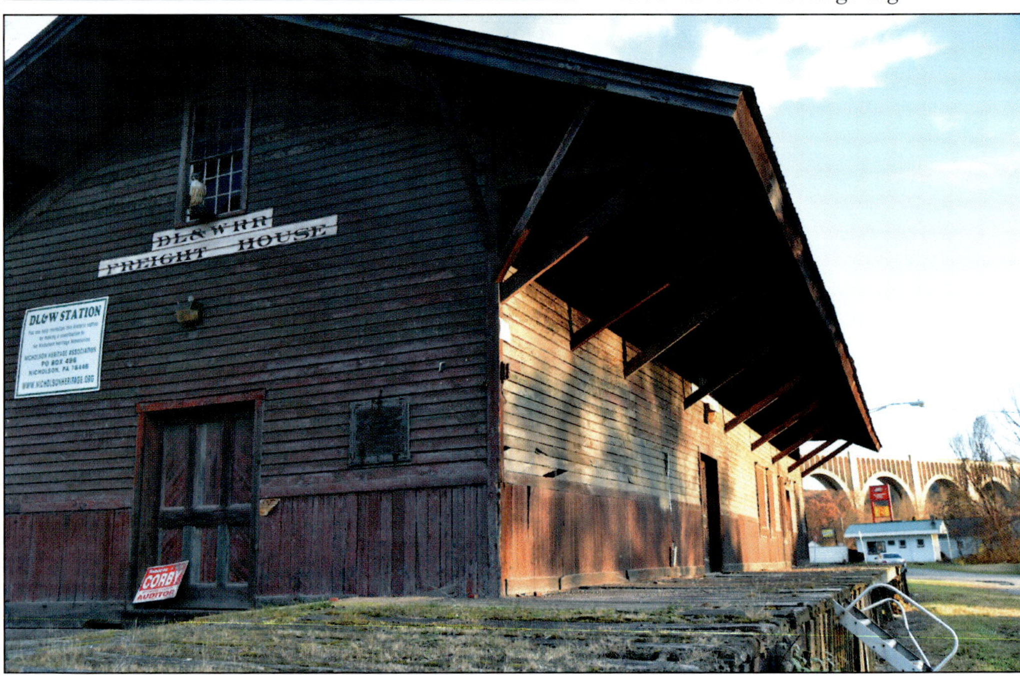

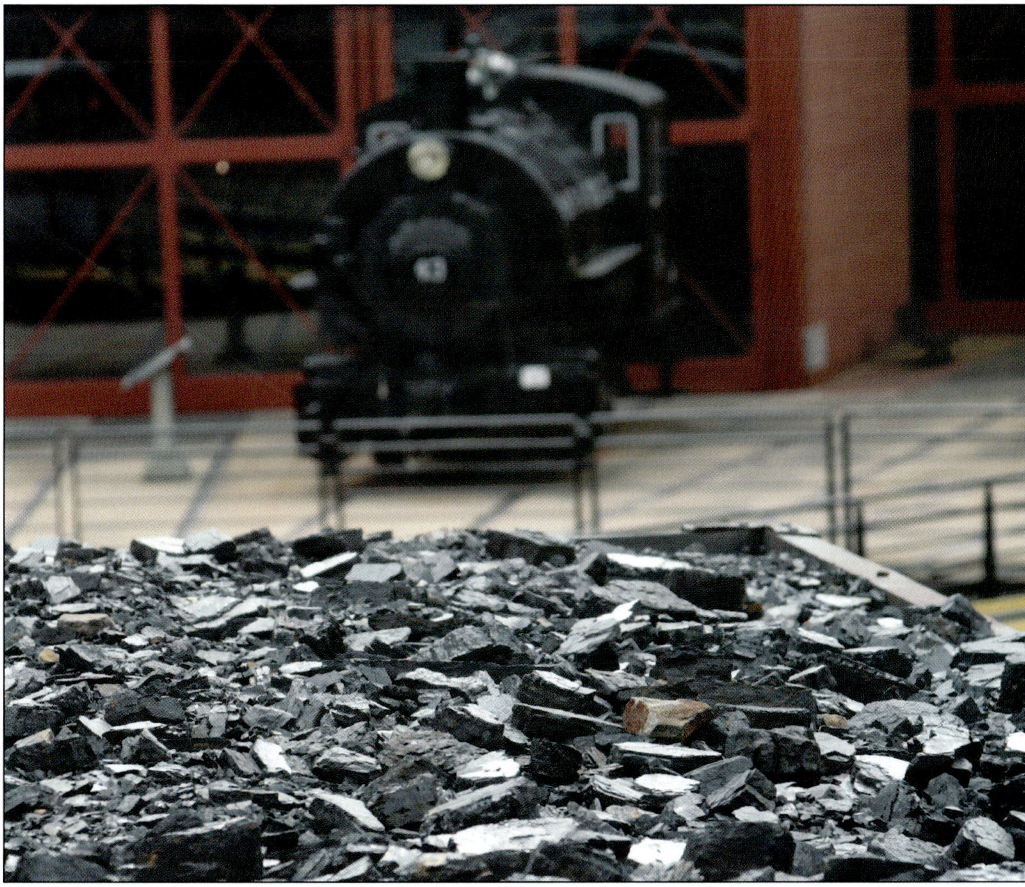

The Steamtown National Historic Site sits upon 40 acres that was once the DL&W's rail yard. The museum is dedicated to educating the public on the significance steam railroads played in the United States' growth during the Industrial Revolution. The site includes a large collection of steam engines and exhibits.

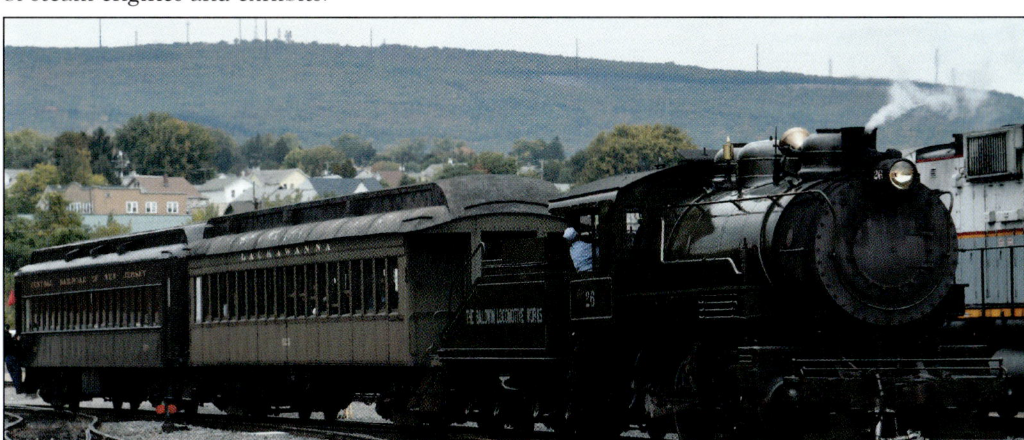

Steamtown also offers rail excursions throughout the year. Destinations vary from local routes through Scranton to more distant forays into the Pocono Mountains. One of the most popular excursions is the Holiday Express during the Christmas season, where riders meet Santa.

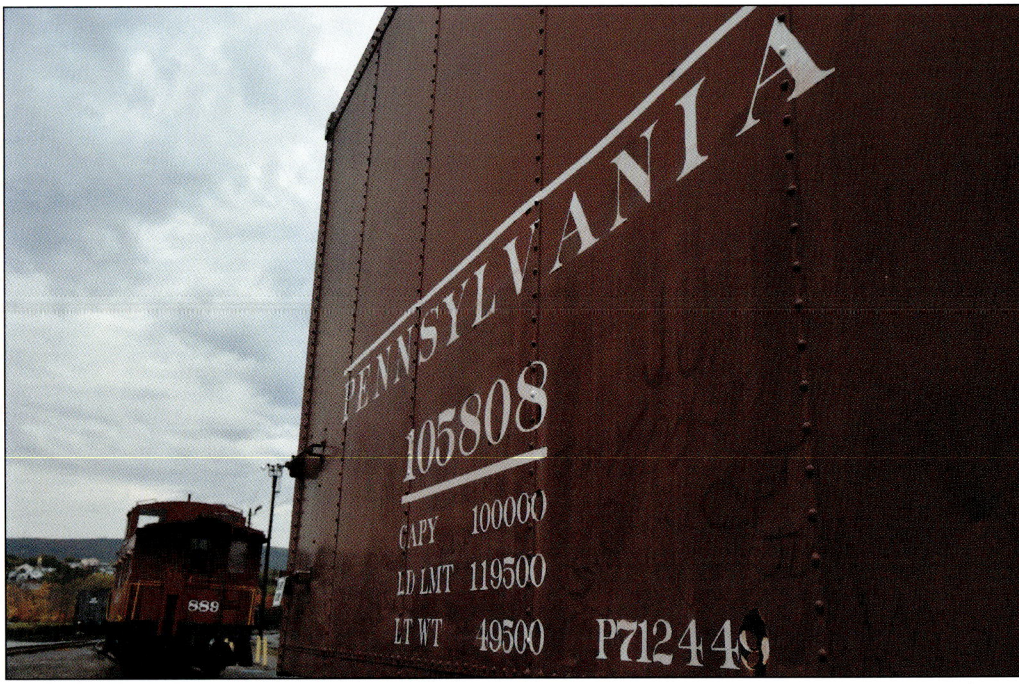

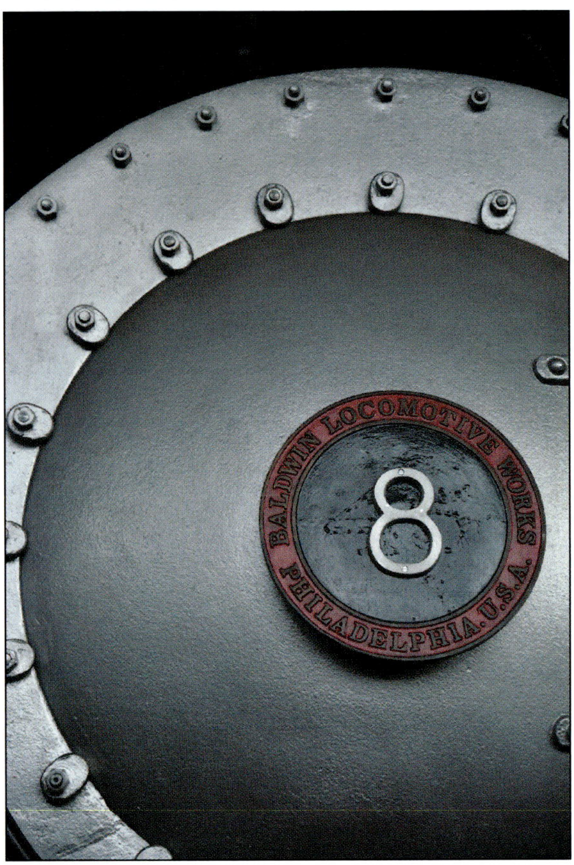

Steamtown's collection of railroad cars and engines was amassed by Nelson Blount. Blount was a railroad owner from New England who was also president of a multimillion dollar corporation. In addition, he coauthored the book *Along the Iron Rails* with Fred Richardson. Blount realized his lifelong dream of a railroad museum when he established Steamtown U.S.A. in Riverside, Vermont. When he died in an aviation accident, the museum soon floundered.

In 1995, Scranton's Steamtown opened to the public after local officials lobbied aggressively to relocate the collection from its original home in Riverside, Vermont. The city competed against 30 other cities to become home to this historic collection of motive power.

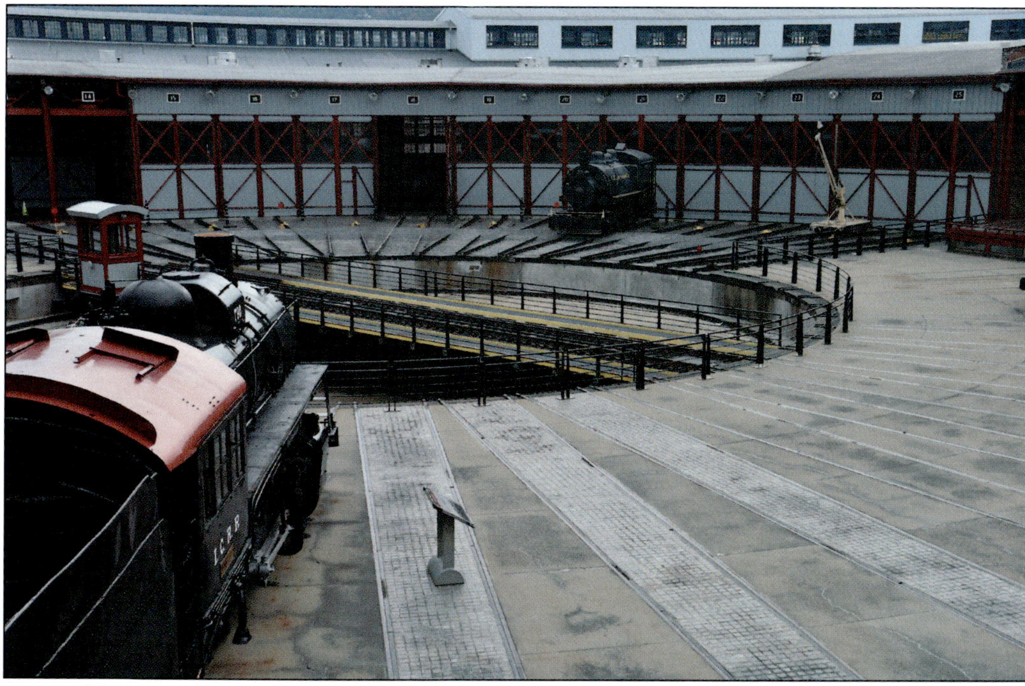

Pictured here is the roundhouse and turntable pit. Both were meticulously restored after years of neglect. The roundhouse now serves as a space to rehabilitate and care for locomotives with light-duty maintenance and repairs. The turntable pit is once again operational as it moves cars in and out of the roundhouse.

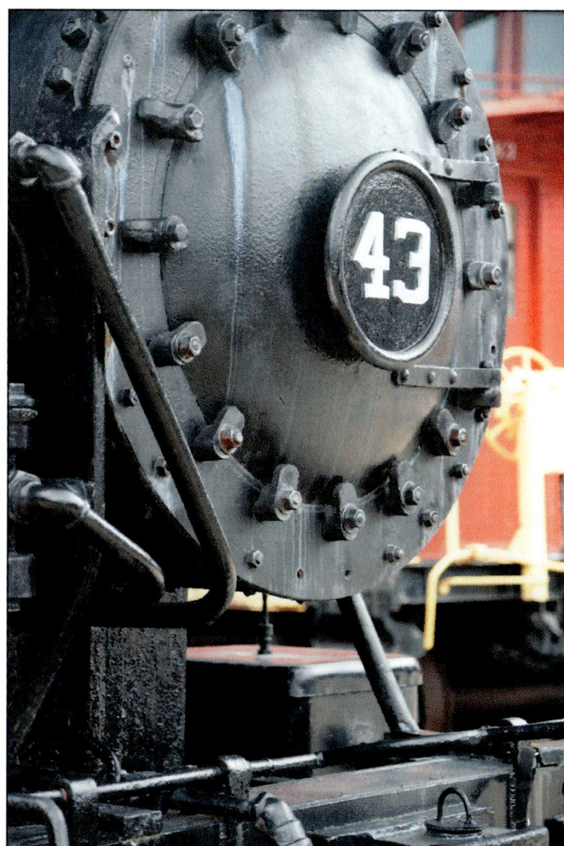

The train collection at Steamtown attracts railfans and regional history buffs. Pictured here is Locomotive No. 43, which satisfies both visitors. In 1918, Vulcan Iron Works in nearby Wilkes-Barre built the saddle-tanker that now sits alongside the turntable pit.

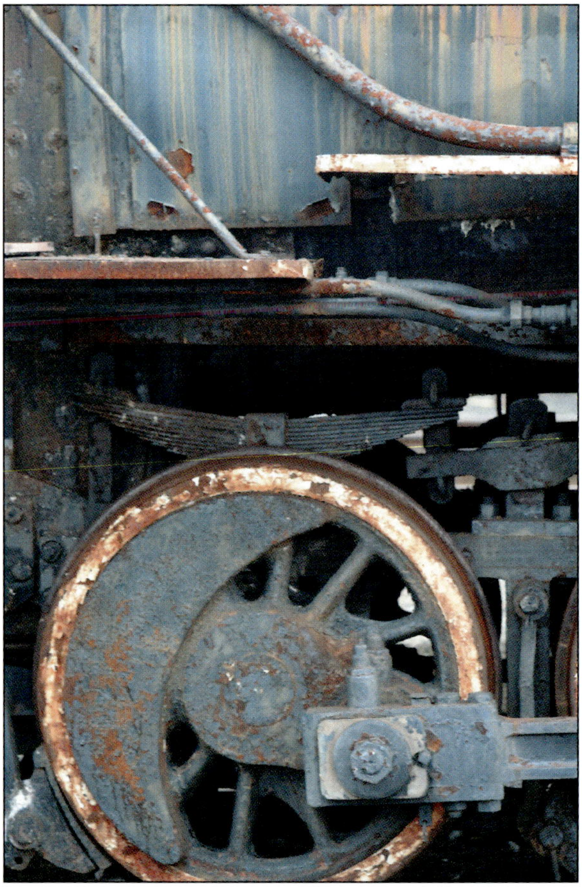

Railroad traffic slowed down significantly by the 1950s, and in some cases, some lines ceased operation all together. Competition from automobiles and trucks winnowed away passengers and freight from the once dominant steam engines. In addition, the introduction of diesel locomotives eventually made steam engines obsolete.

The DL&W finally ceased existence in 1960. In a last-ditch effort to save the company, the DL&W merged with its competitor the Erie Railroad. What resulted was the creation of the Erie Lackawanna Railway (later Railroad). Unfortunately, the railroad only lasted until 1976, when Conrail acquired it.

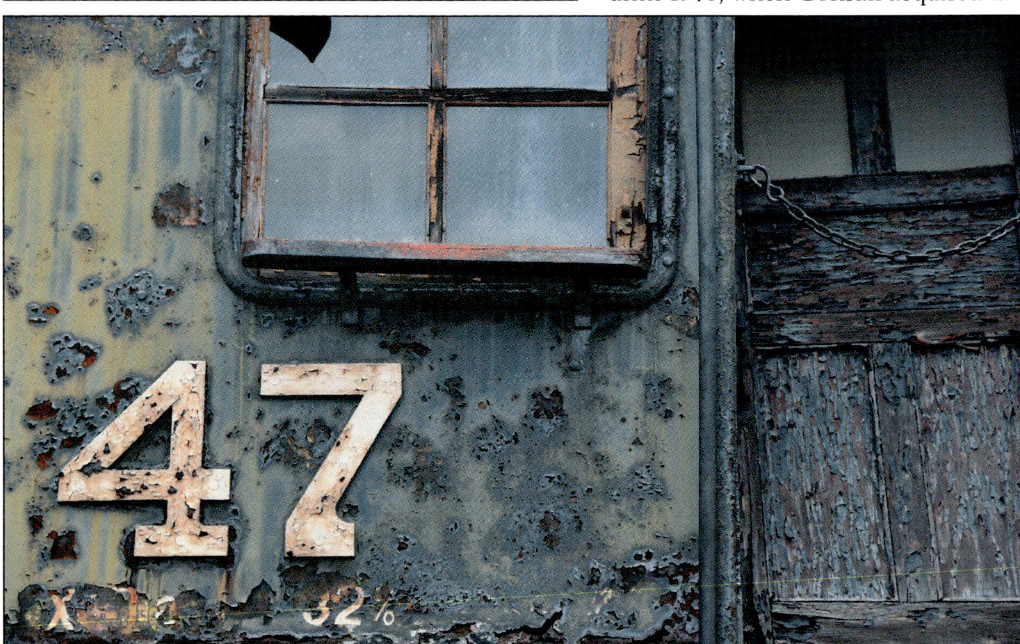

Five

Centralia

When tourists visit Centralia, they expect to see a town shrouded in smoke with flames rising up from the earth. It is a reasonable expectation since the borough sits upon an underground mine fire that has been burning for over 50 years. What visitors actually see are blocks of paved streets with no homes on them. However, even on cold days in December, it is hard to find vents that yield much visible steam. Were it not for the handful of residents that still live here, Centralia would be a ghost town. Though the fires may be out of sight, they are there to stay—for perhaps another one or two centuries if no method of extinguishing them is found.

The State of Pennsylvania is working on putting out mine fires but unfortunately not in Centralia. In 2016, Gov. Tom Wolf announced that $30 million of federal funding would be used to reclaim abandoned mine lands. Although Centralia did not make the list, one area in coal country did.

Burning for over two decades, the Powderly Creek underground mine fire is located just outside the city of Carbondale. The federal money used to squash it is part of a pilot program. It is hoped that once the land is reclaimed, it could be used for development.

For now, the state's most famous mine fire rages on in Centralia and will do so into the unforeseen future.

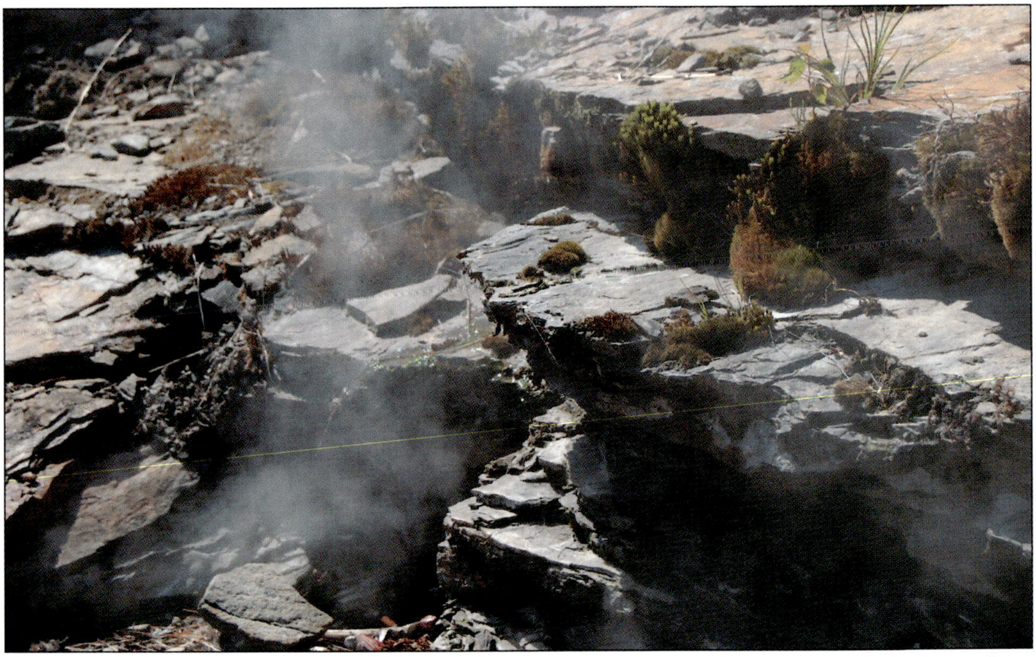

One of Centralia's biggest mysteries is how the fire started. There are a few theories as to how it ignited, but there is no definitive evidence as to who started it. Most theories involve one common thread: trash was allegedly burned too close to a coal vein and ignited the coal beneath. (Courtesy of Kelly Mifsud.)

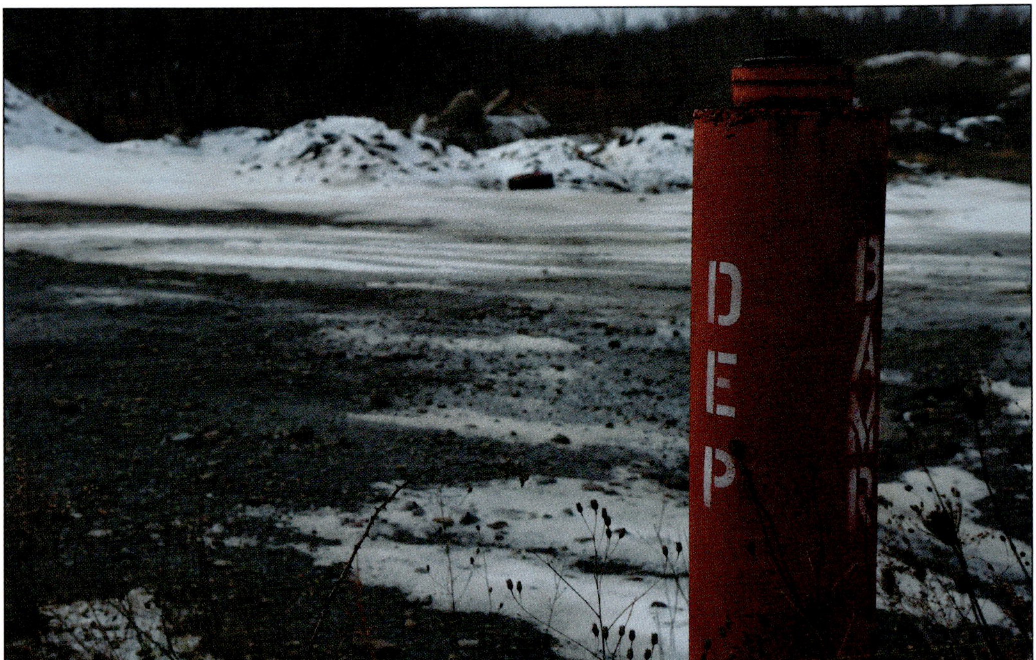

In 1962, the ground beneath Centralia started burning beneath its population of over 1,200 people. Attempts were made to squelch the fires but none were successful. Pictured here is a pipe placed by the Department of Environmental Protection and the Bureau of Abandoned Mine Research.

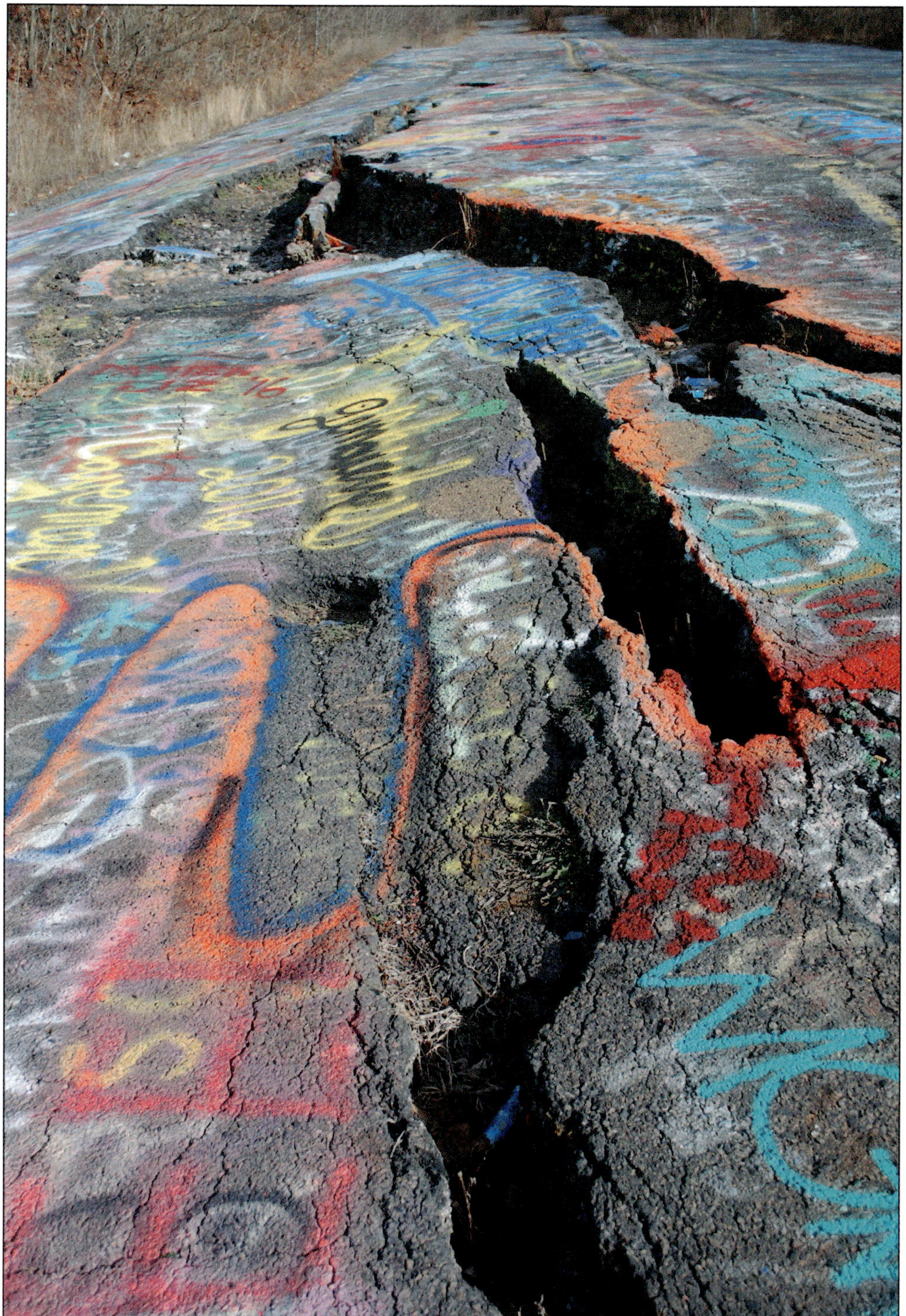

Now known as "Graffiti Highway," this stretch of route 61 was closed to traffic when it started buckling from the flames underneath it. Seen here, it appears an earthquake split the road apart. The highway has become such a magnet for tourists that it is now included on Google Maps.

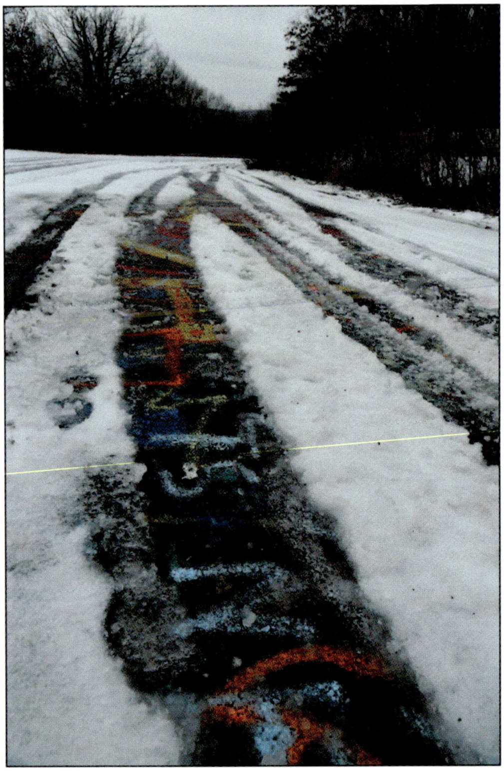

Route 61 was rerouted in 1994, leaving this abandoned highway to off-roaders and tourists. One couple that the author encountered exploring Graffiti Highway said they were a little disappointed there was no visible steam. That is because the fires have moved to another part of the borough and continue to move as they consume the coal beneath.

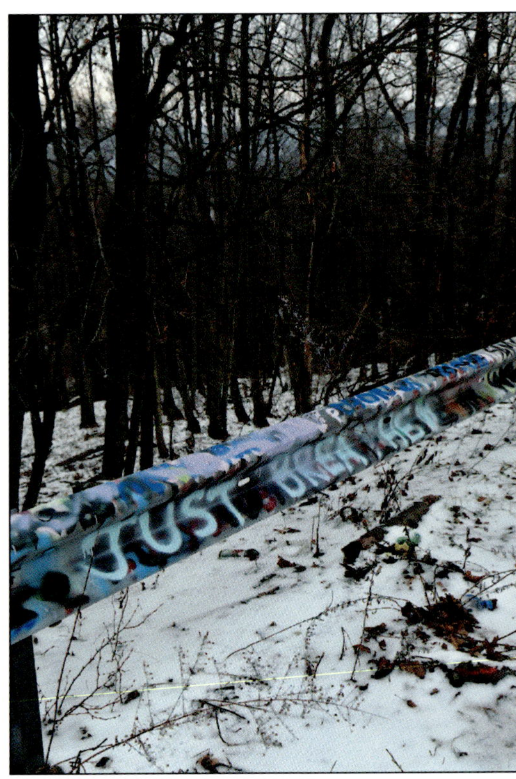

This tagger's message is a very dangerous one. The fumes that rise through the air can be quite dangerous to inhale. The carbon dioxide emissions became a great concern to residents when the fumes started to infiltrate the basements of homes and businesses. One gasoline station owner became alarmed when his underground gas tank's temperature rose dramatically.

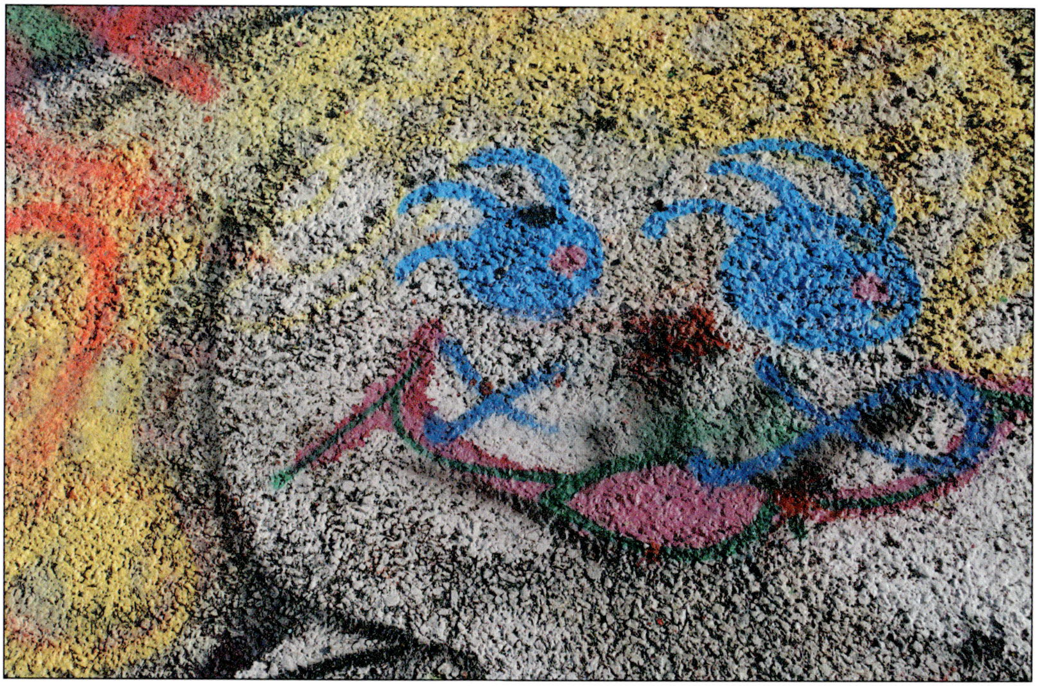

While Graffiti Highway has been blanketed with an alarming amount of phallic imagery, there are many spots where taggers have attempted to be a bit more creative. Over the years, layers of tagging have made this an evolving public artwork piece.

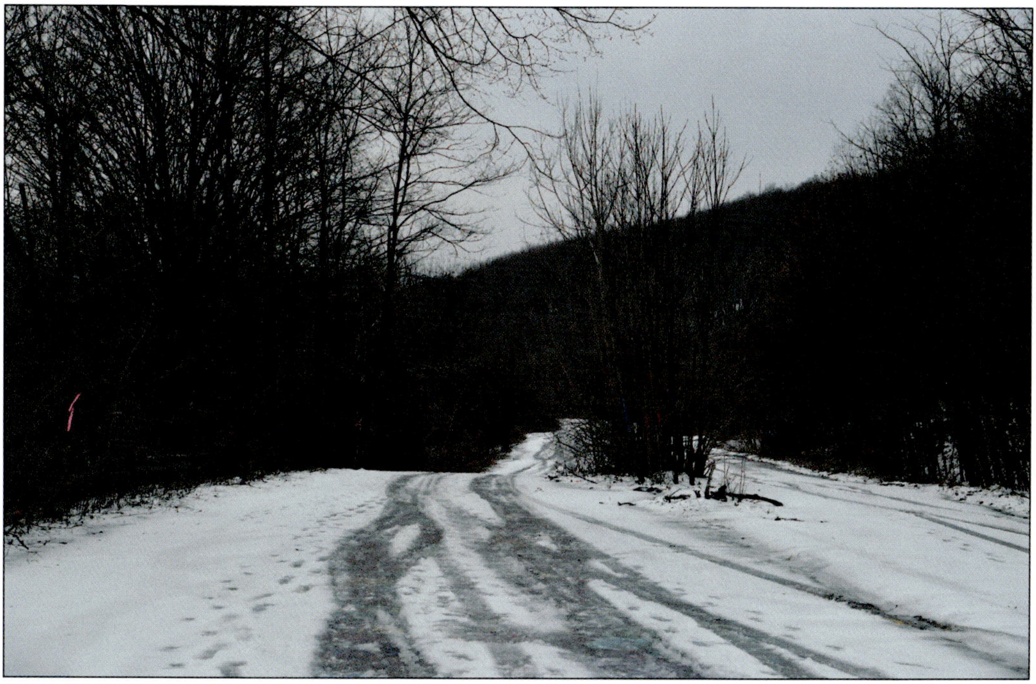

Evidence that underground mine fires are not as active here is evident by the amount of snow sticking to the road in this picture. Other areas of Centralia have reported snow melting rapidly on roads because of the smoldering fires below.

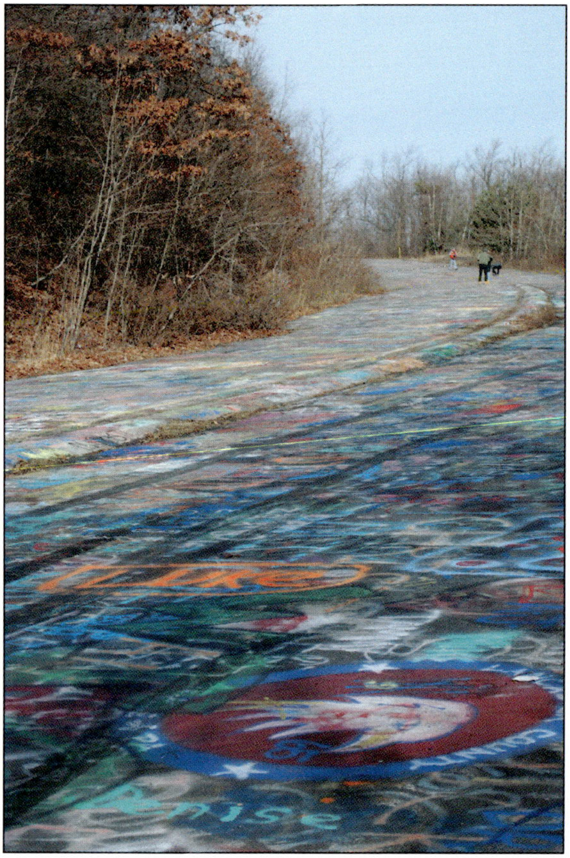

Also known as the "Highway of Death," Route 61 should not be trodden upon without caution. The Department of Environmental Protection has warned that walking or driving in the area can result in serious injury, or death, since the road could collapse.

In 1984, Congress appropriated over $42 million to relocate Centralians. Most of the residents accepted the buyout offer and moved to nearby boroughs. However, there were some holdouts who did not wish to move. Further pressure was put on them to relocate in 1992. At that time, the state invoked eminent domain and condemned all the buildings in the borough.

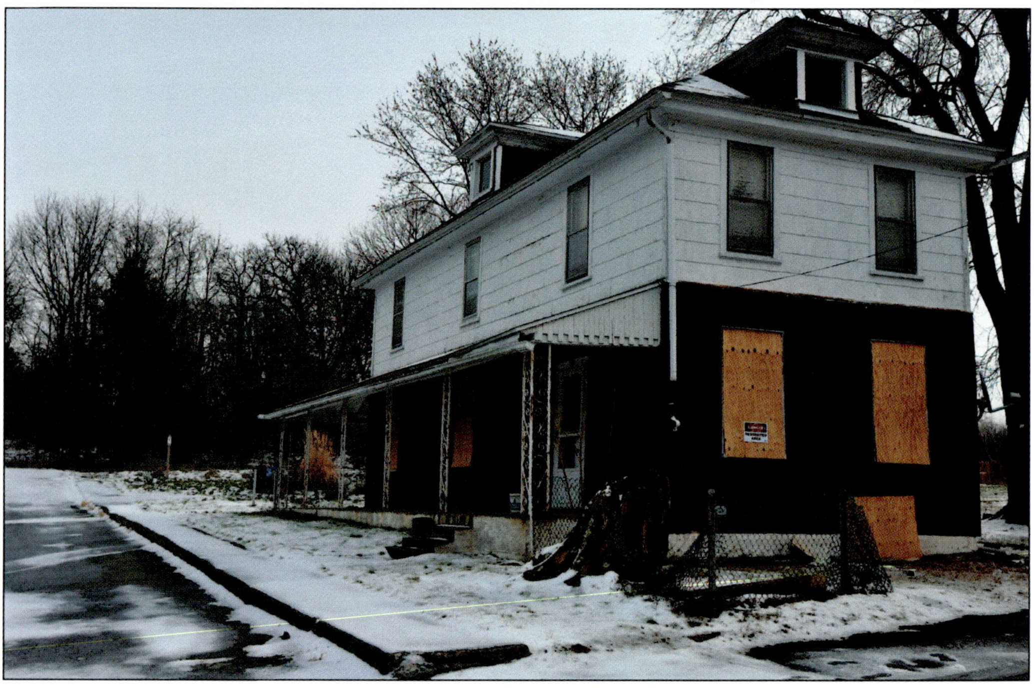

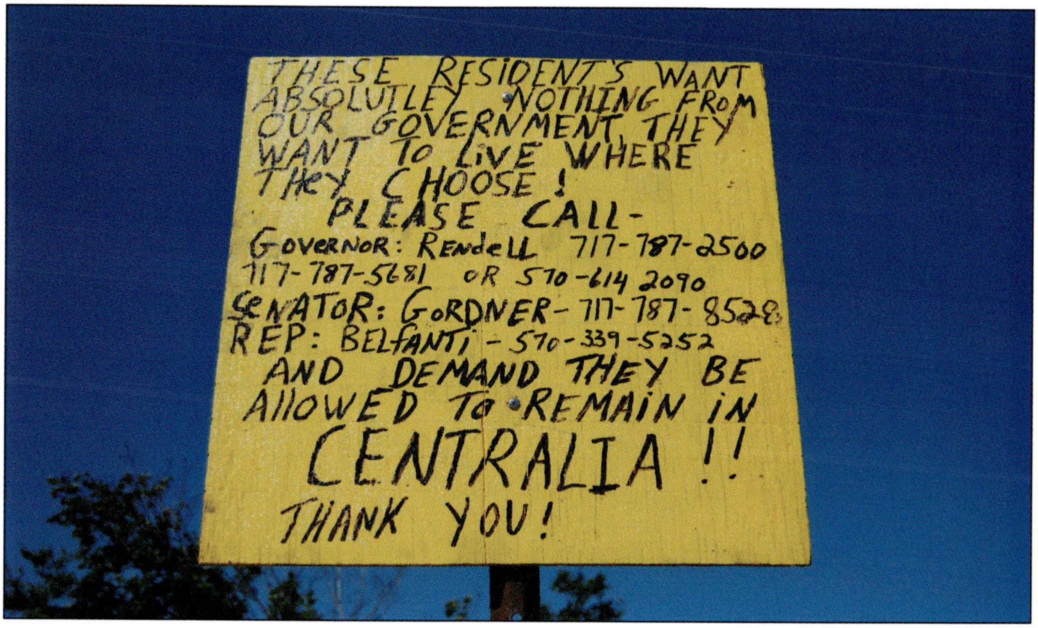

In 2009, Pennsylvania governor Ed Rendell embarked on the eviction of the remaining residents. Residents were not silent about their opposition to their eviction. In addition to posting these signs, Centralians turned to local and national media to state their case. At least one resident voiced the theory that the government's ouster of residents was part of a conspiracy to acquire the property that still held tons of valuable coal beneath. (Courtesy of Kelly Mifsud.)

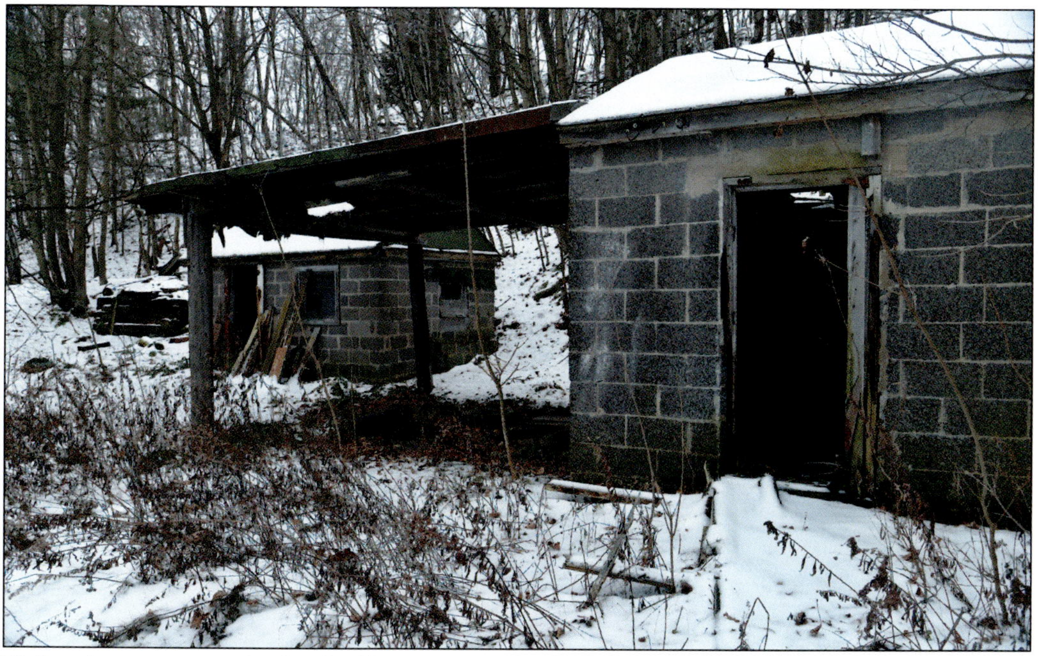

Today, there are more derelict buildings remaining than there are occupied homes. Even for those homes without No Trespassing signs, it is prudent to avoid entering these structures because they can pose a myriad of hazards. In addition, the land leading to vacant structures may contain holes that are not visible because of debris buildup or snow.

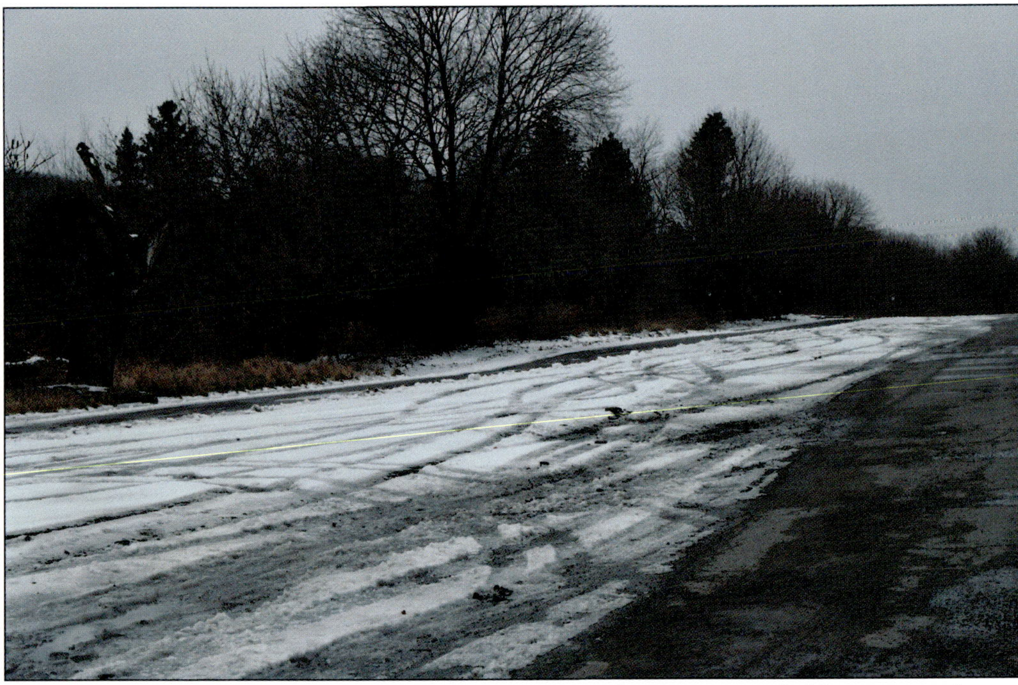

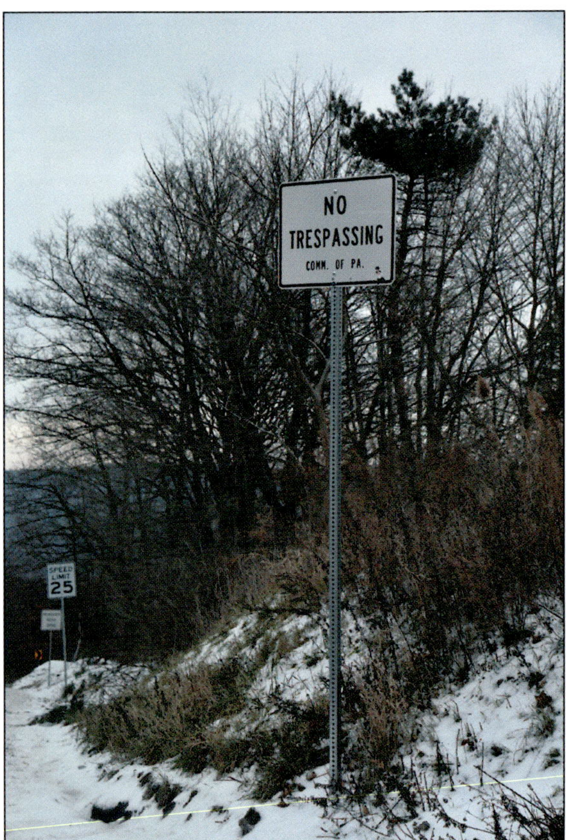

The ghost-town feel has attracted filmmakers, television production, and journalists from around the world. The movie *Silent Hill*, which is based on a video game, was inspired by Centralia. The film was released in 2006, but the film's crew never shot in the borough. Instead, the filmmakers chose Canada as their location.

This No Trespassing sign reminds visitors it is illegal to tread upon this land that is now owned by the US government. This sign is located at the entrance to Graffiti Highway, the old Route 61.

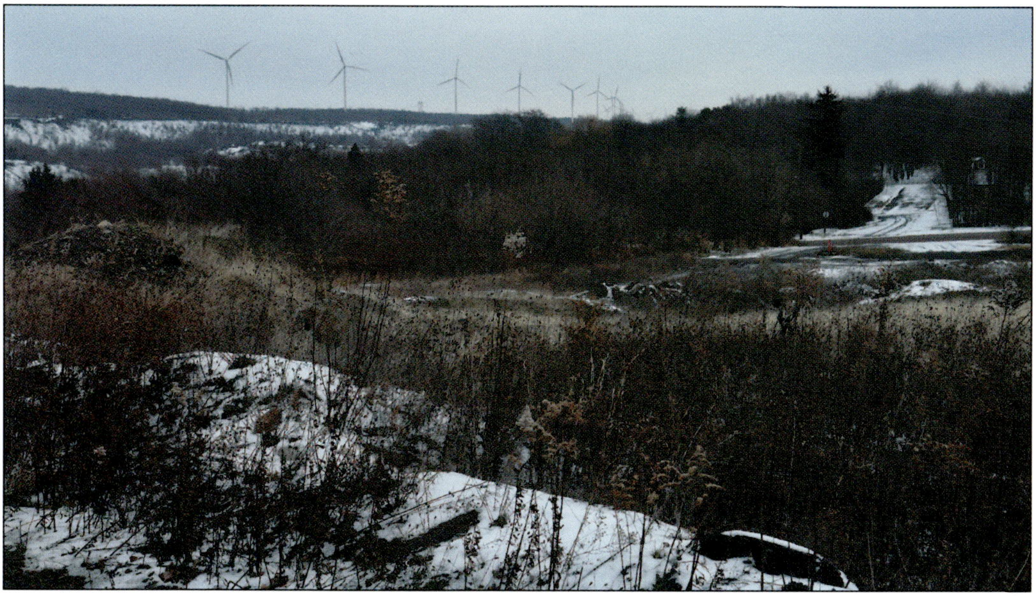

Centralia's rich coal deposits brought wealth to the area in the 1800s but, unfortunately, led to its demise. Even before the fires raged beneath, there were a multitude of problems that came with the coal industry's arrival in the borough. From the murder of the town's founder by the Molly Maguires to unsafe mining condition, the town had more than its share of challenges.

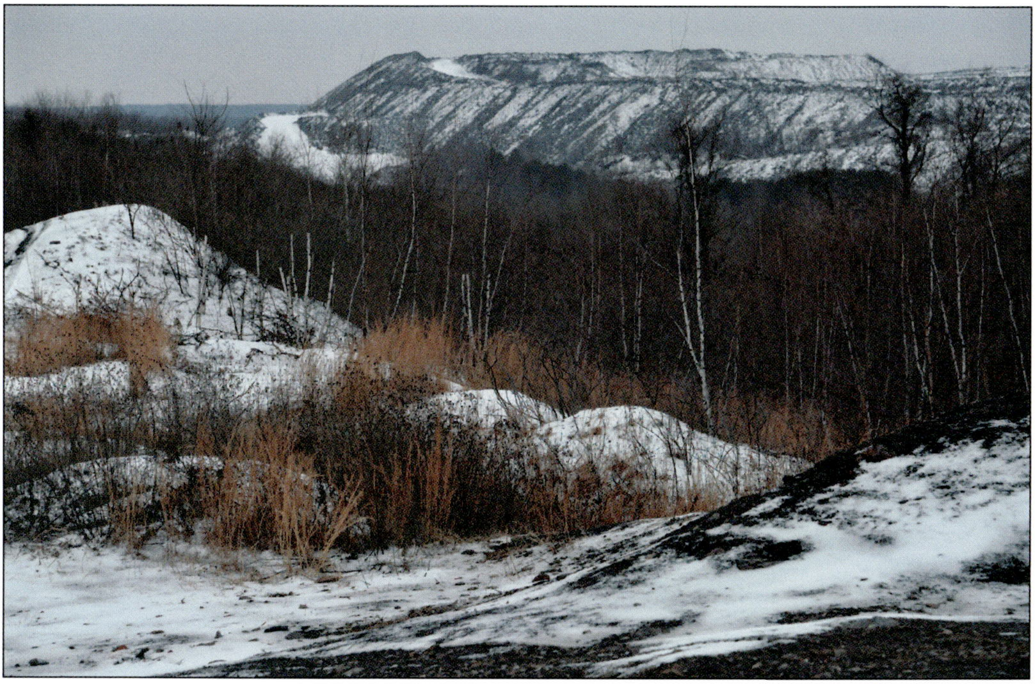

Culm banks surrounding Centralia hint at the wealth of coal underground. That wealth attracted the town's founder, Alexander Rea, who was a mining engineer. One night while traveling home from neighboring Mount Carmel, he was murdered in his buggy by men from the infamous Molly Maguires. While it took almost a decade to convict the men for his murder, they faced justice when they were hanged.

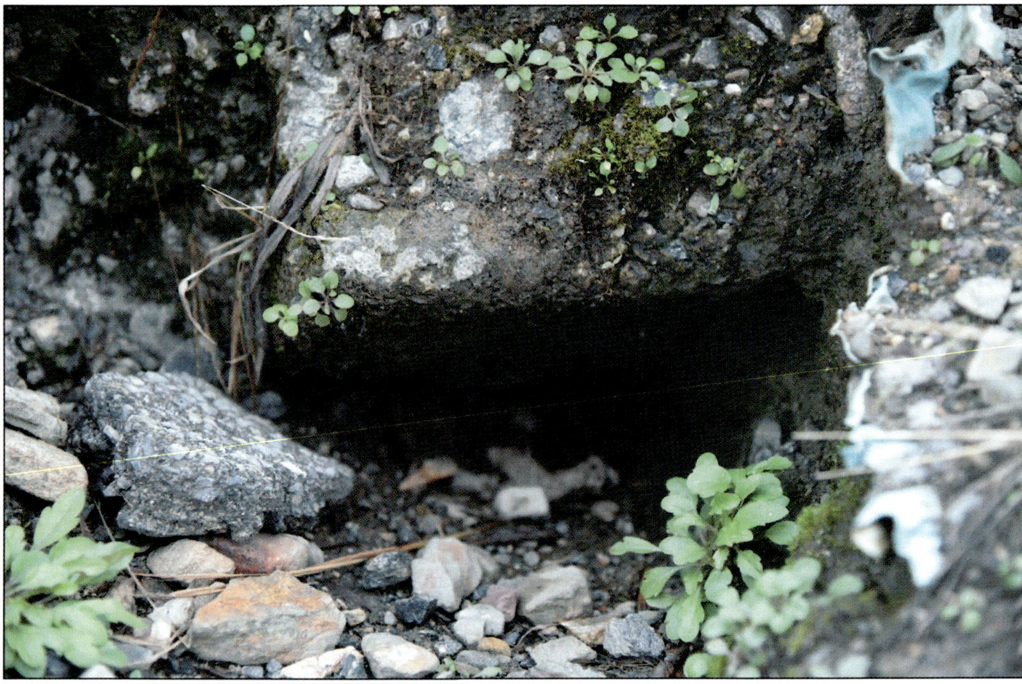

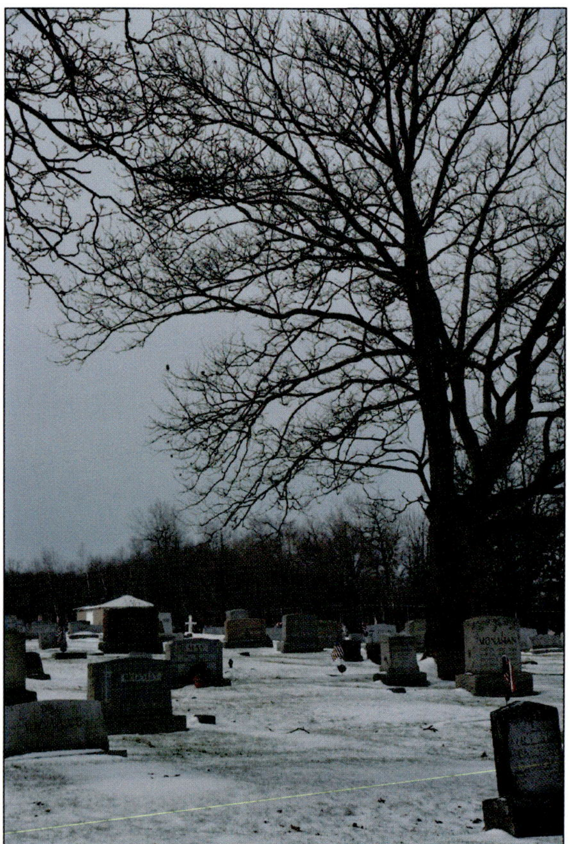

Small vents like this emit steam from the blaze below. As one can see here, the steam is not visible. To best photograph vents, people should visit in winter and, if possible, on a rainy day. Vents with surrounding small plants are a sign there is warmth below.

Pictured here is St. Ignatius Cemetery, one of three cemeteries left in the borough. It is the closest cemetery to where the underground fires first started. As one faces the cemetery, the entrance to Graffiti Highway is to the left.

The graveyard is populated by gravestones marking the passing of many townspeople who never lived to see the fires of Centralia. However, it is reported their graves did. At one point, fires raged beneath St. Ignatius Cemetery, with steam rising from the graves.

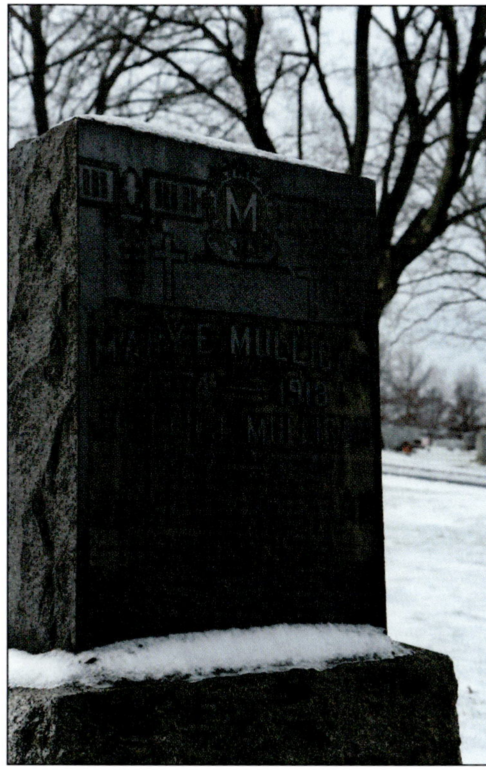

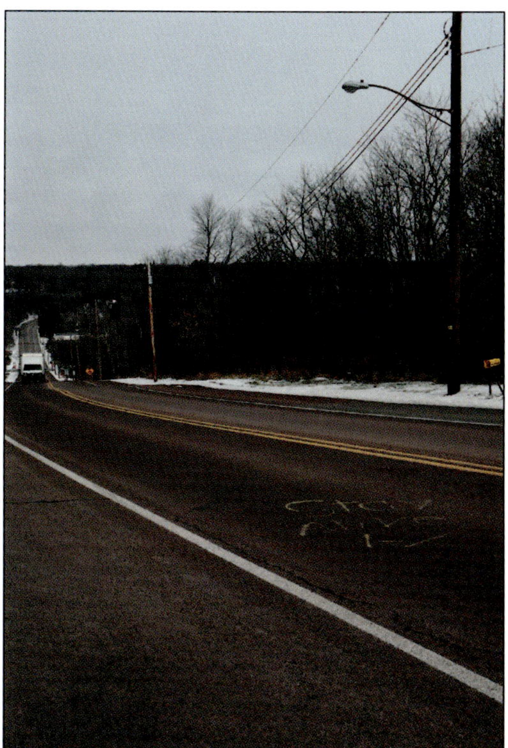

The roads of Centralia today give no hint of the vibrant enclave it once was. Its population peaked at close to 3,000 people in the late 1800s. Streets were lined with 27 bars, 14 stores, 7 churches, and 1 bank.

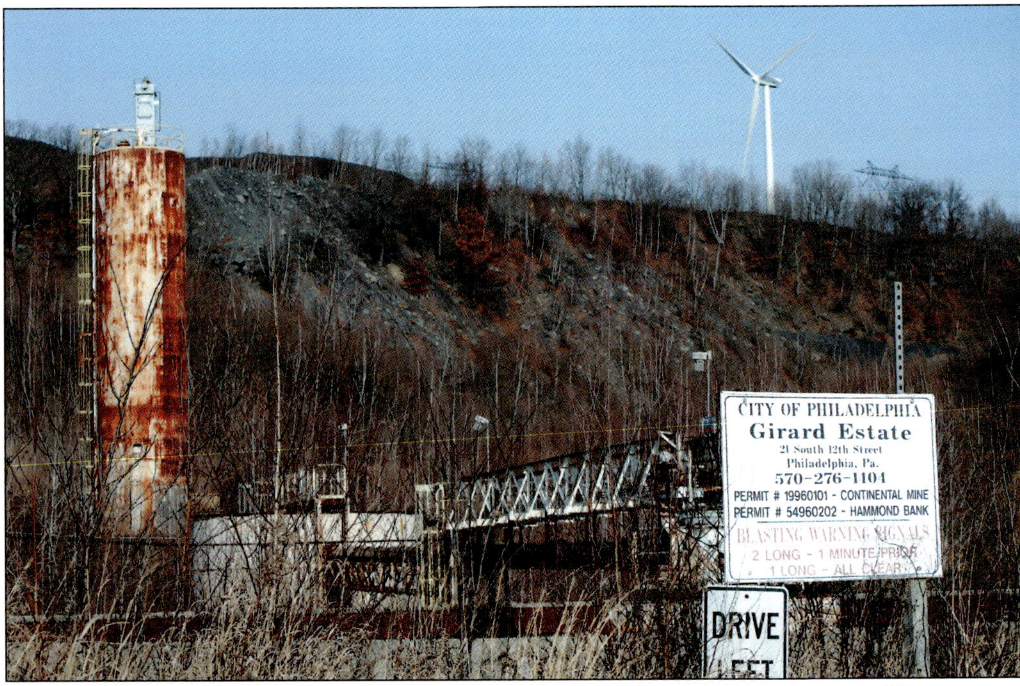

In 1830, Stephen Girard purchased the land Centralia sits upon in an auction. He suspected anthracite coal lay beneath and confirmed his suspicions when he surveyed it later. Girard was a former French mariner turned Philadelphian and was one of the richest men in the country at the time.

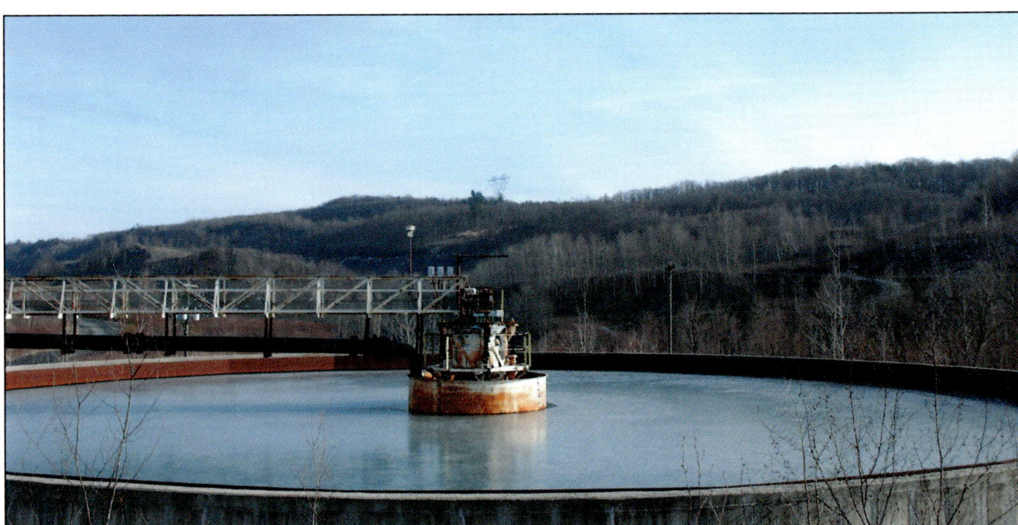

Seen here is a Dorr thickener that was once part of the Continental Colliery on Girard's land. Today, this land is just a small part of Girard's wealth. After his death in 1831, his holdings were put into a trust. A majority of that trust was bequeathed to Philadelphia and New Orleans (his home before Philadelphia). His investment in anthracite coal fueled the trust's growth long after his death.

Two lawn ornaments stand in the front yard of a home that has long been demolished along Route 61. Throughout the borough, fire hydrants, stop signs, and sidewalks populate the area, reminding visitors these usually commonplace objects no longer serve a purpose.

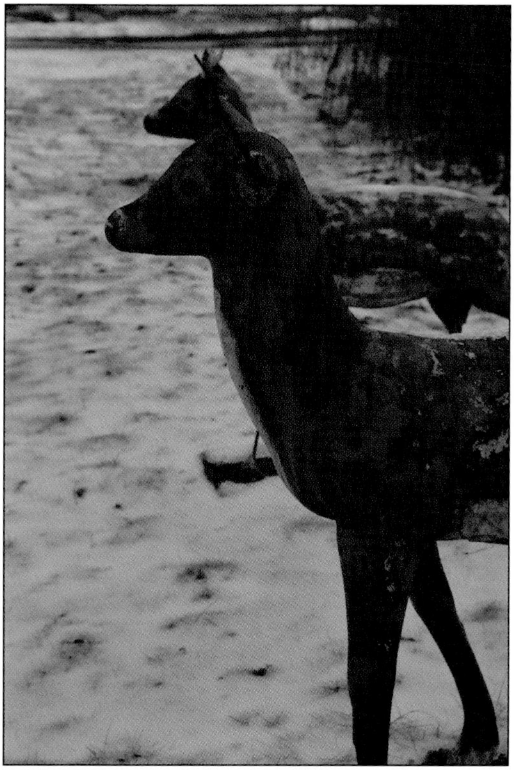

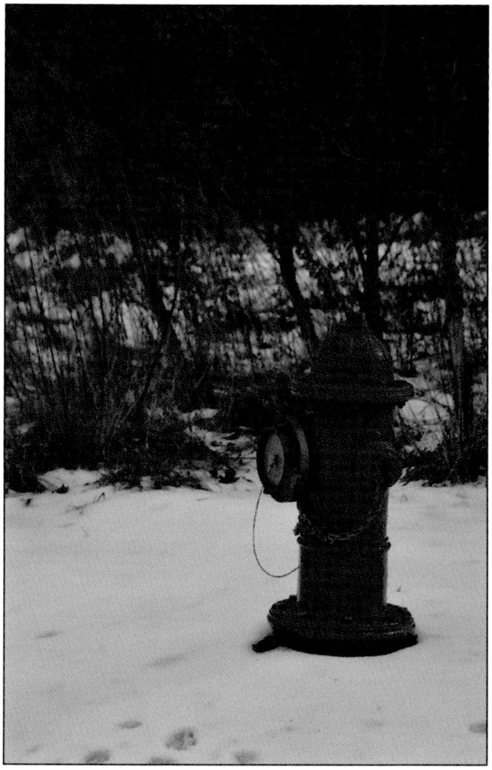

While Centralia's flames may have thwarted daily life, they have yet to claim a life. There was a close call though when 12-year-old Todd Domboski fell into a sinkhole in 1981. His cousin Eric Wolfgang heard his cries for help and rescued him from the smoking pit.

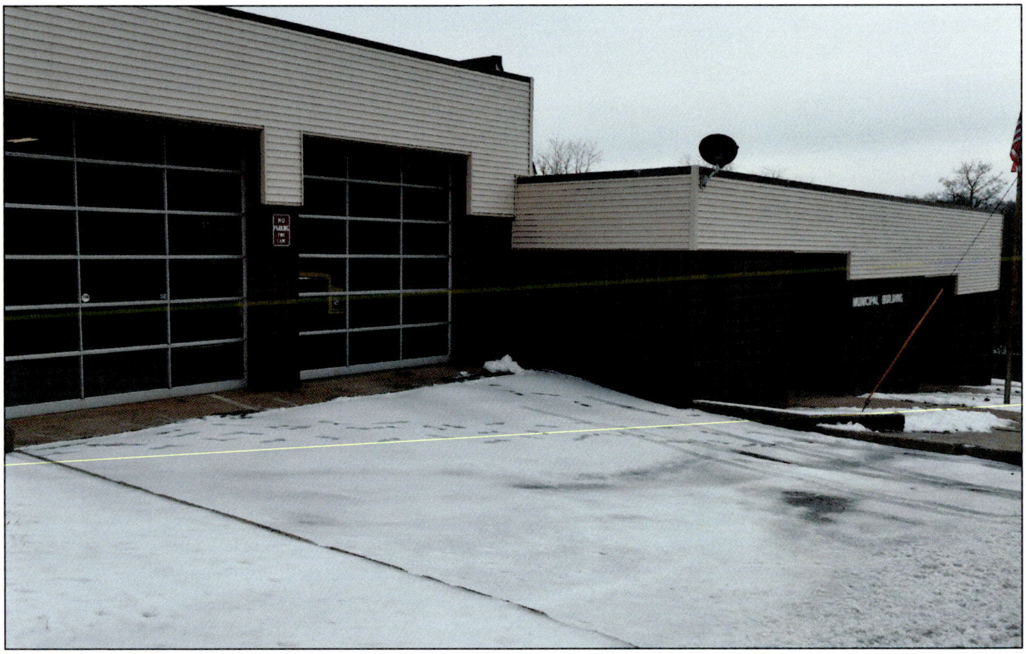

In 1978, Centralia's destiny was decided here in this municipal building. Residents voted to determine if they should shutter the borough or stay among the flames that had already been burning for 16 years. Those in favor of leaving won the vote, but it was not a unanimous decision.

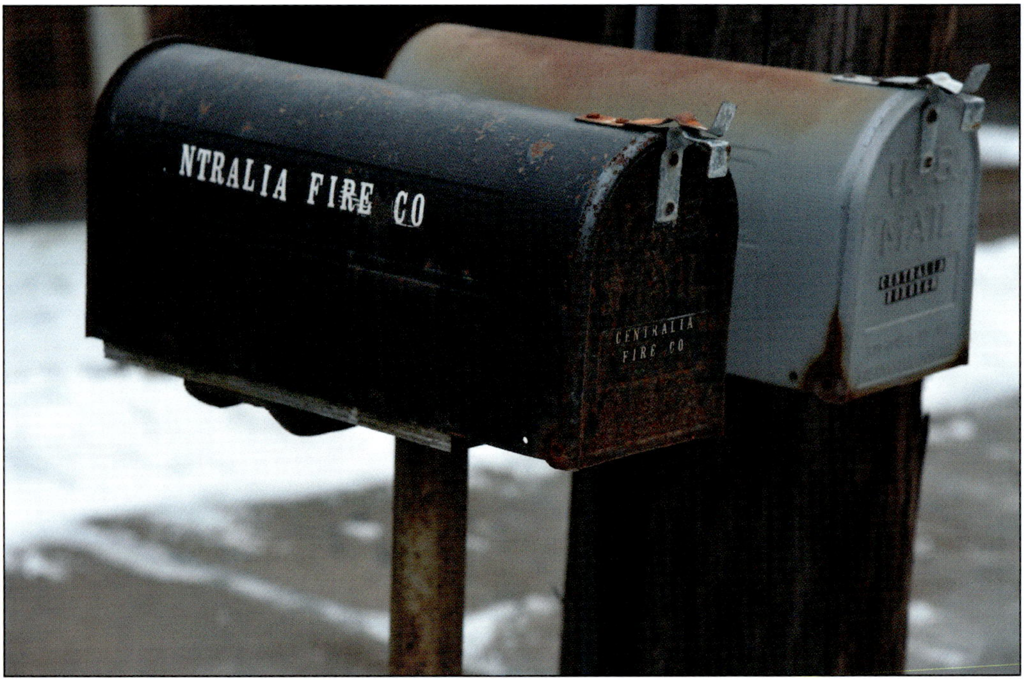

By 1990, there were only 63 Centralians left. In yet another blow to the remaining residents, Centralia's post office was demolished in 1997. With the dwindling population, mail was less frequent. Then in 2002, the US Postal Service officially revoked the town's zip code, 17927.

Six

Coal Patches

Once the long workday ended, it was time to head home from the breaker or the mines. But for many, home and work were intertwined, since their rentals were owned by the coal company. To make matters worse, homes in the coal patch were crude and cramped because money was in short supply for better housing.

Eckley Miners' Village Museum is a glimpse into how coal miners, laborers, and their families lived in the late 1800s. Located in Weatherly, this indoor/outdoor museum is operated by the Pennsylvania Historical Museum and Commission. From the Industrial Revolution to, surprisingly, Hollywood history, the site has a plethora of stories to share.

Perhaps the most profound experience Eckley offers is the chance to feel gratitude. "[Visitors] are amazed that people lived like this. The realization makes people thankful for what they have and the sacrifices their forefathers made," says Eckley Miners' Village interim site administrator Dr. Bode Morin.

With the site's 90 acres to explore along with a visitors' center with exhibits, guests should wear comfortable shoes and take plenty of time to walk back in time.

Photographs in this chapter were taken at Eckley Miners' Village Museum or Concrete City.

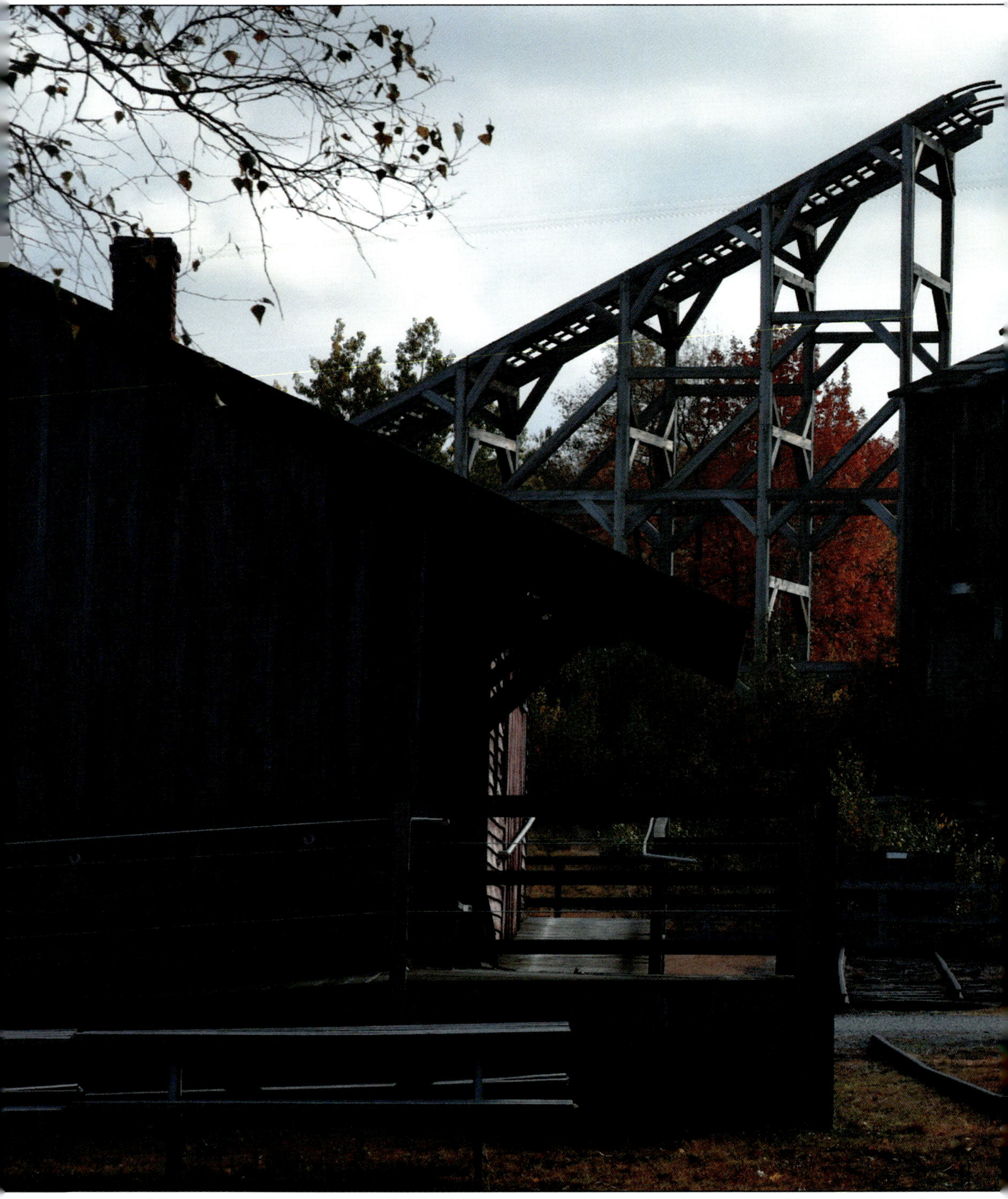

One of the most striking sights in Eckley Village is something that looks quite like a coal breaker but, in fact, is a piece of Hollywood history. The breaker shown here was built for the filming of the 1970 movie *The Molly Maguires*, which was produced by Paramount Pictures and stars Sean Connery and Richard Harris. The village was chosen as a filming location because it looked

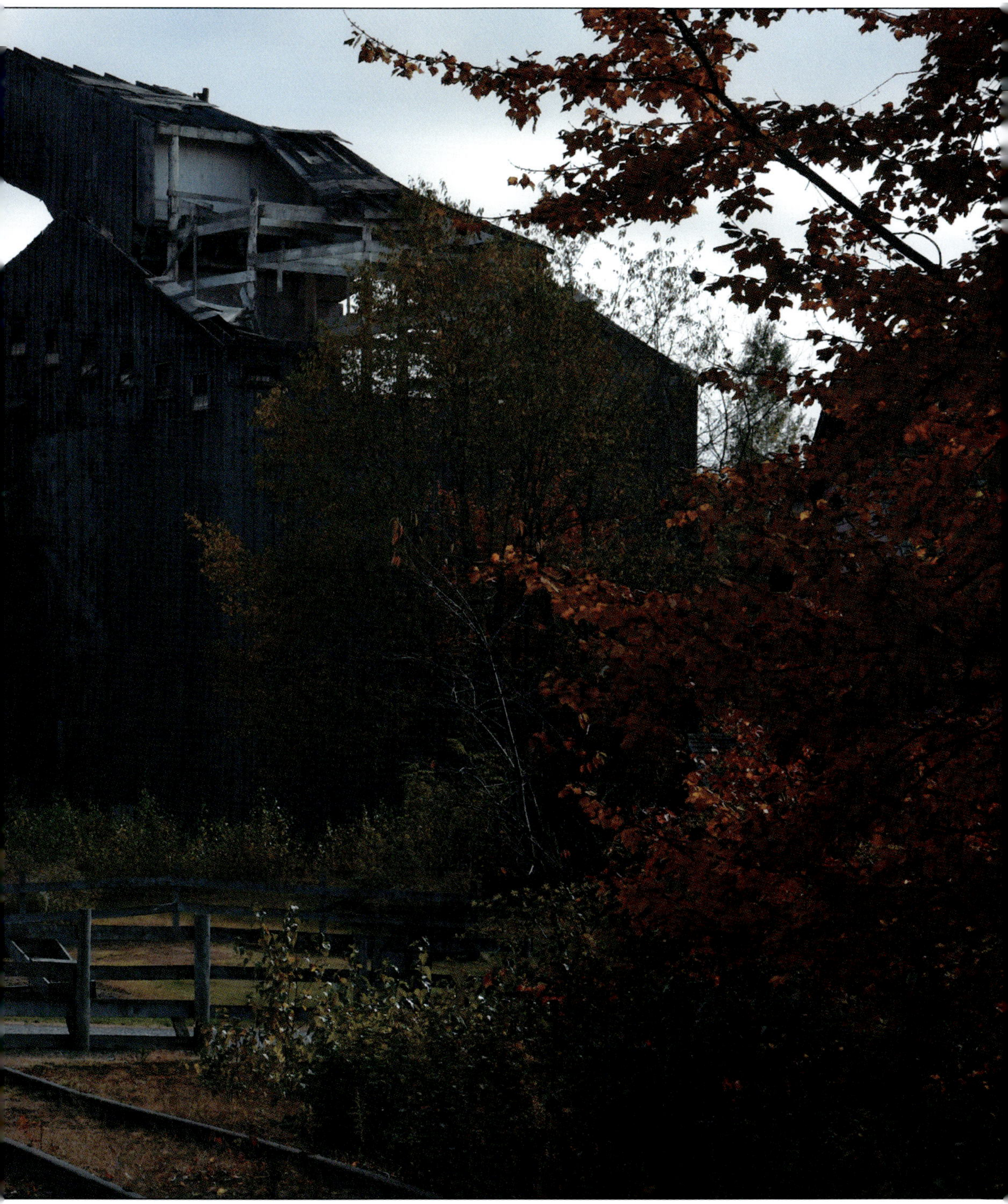

much the same as it did in the late 1800s. This breaker was crucial to the movie's plot and is prominently featured in the film. For a quick look at the breaker's starring role, readers can check out the movie trailer on YouTube.

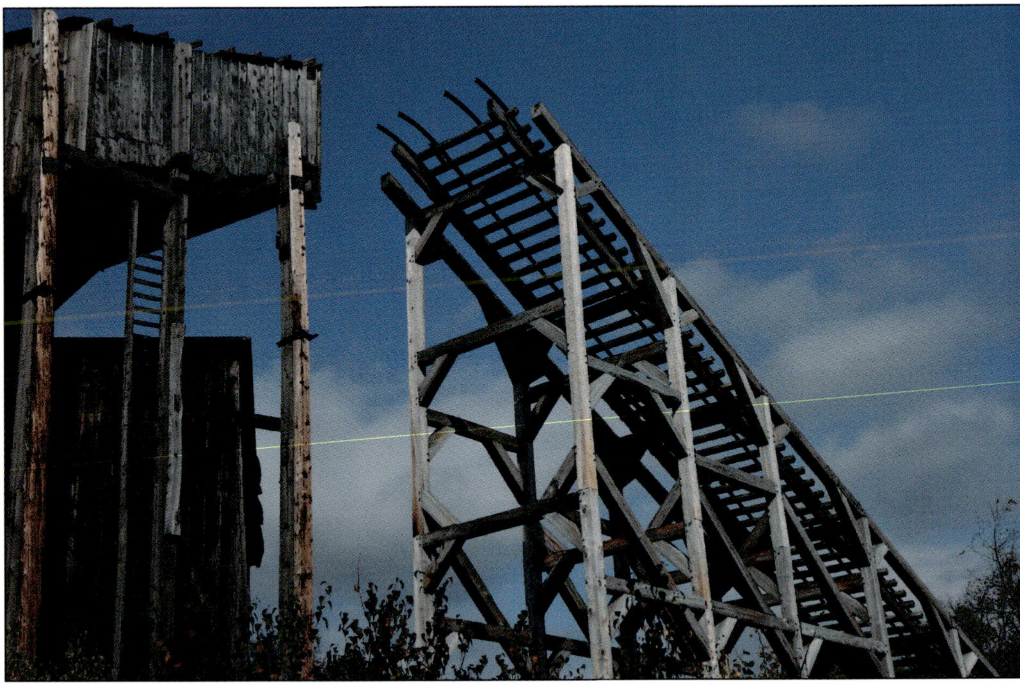

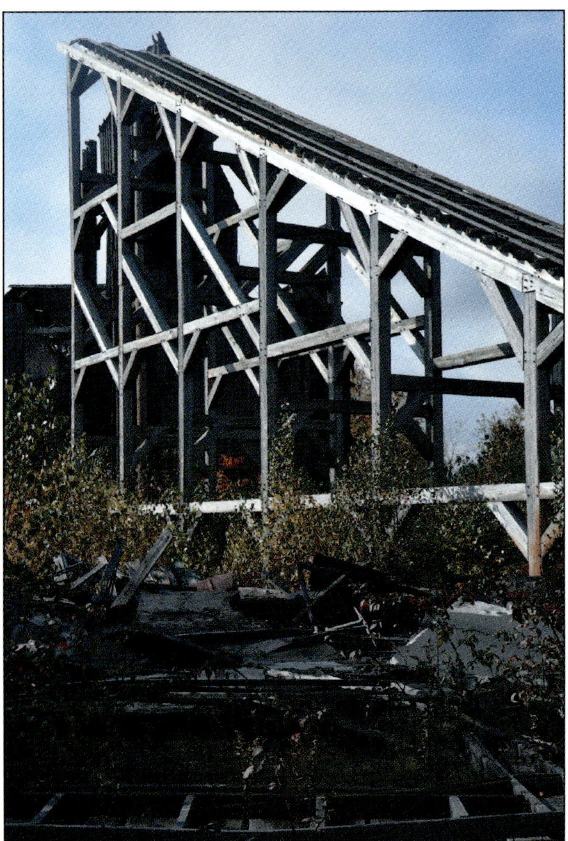

Standing for close to 50 years, the wooden movie set has been crumbling for decades. According to Morin, "There are no preservation or demolition plans for the breaker."

The Molly Maguires is adapted from a 1969 novel by Arthur H. Lewis titled *Lament for the Molly Maguires*. The book was partly inspired by true events that were driven by the secret society comprised of Irish immigrants. History has viewed them as either murderers or heroes for the workingman.

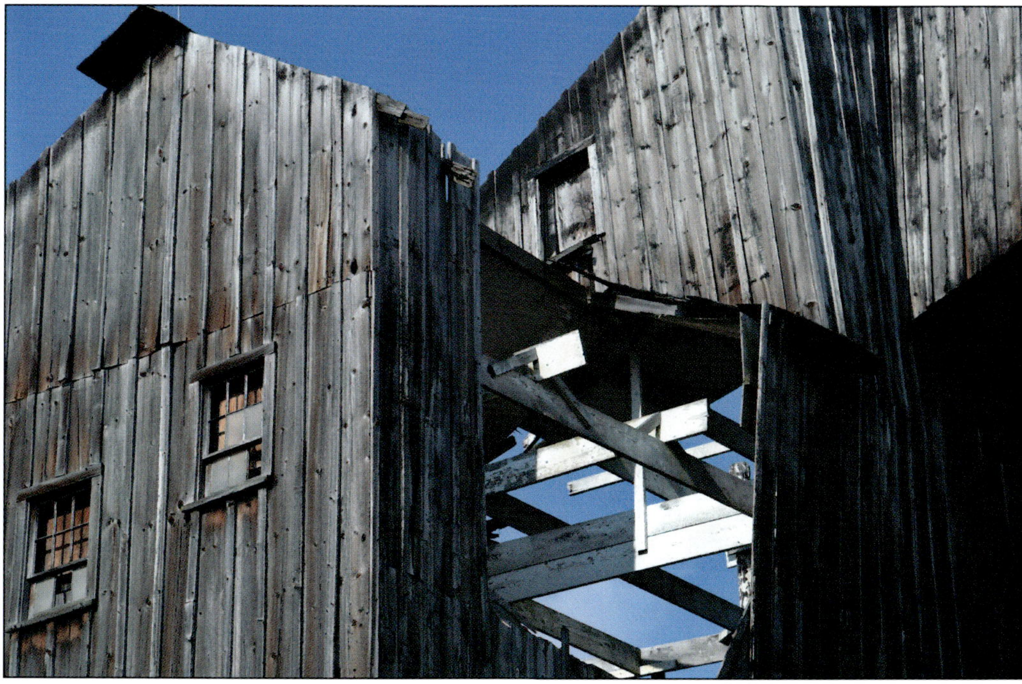

This piece of Hollywood history is responsible for actually saving American history. Because of *The Molly Maguires* production here, state officials decided it was time to save this coal patch from demolition. In 1970, the Pennsylvania Historical Museum and Commission established the historic site.

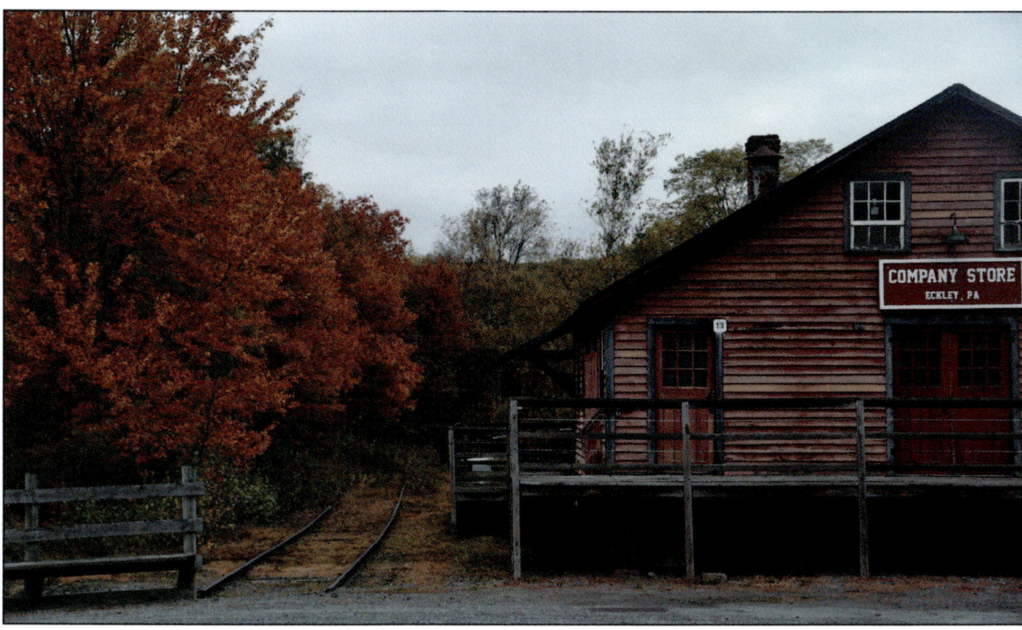

The company store pictured here was also built as part of the set for *The Molly Maguires*. Company stores were owned by coal companies, which often overcharged workers for goods and required that workers make purchases there. The store sold food, mining equipment, clothes, and penny candy.

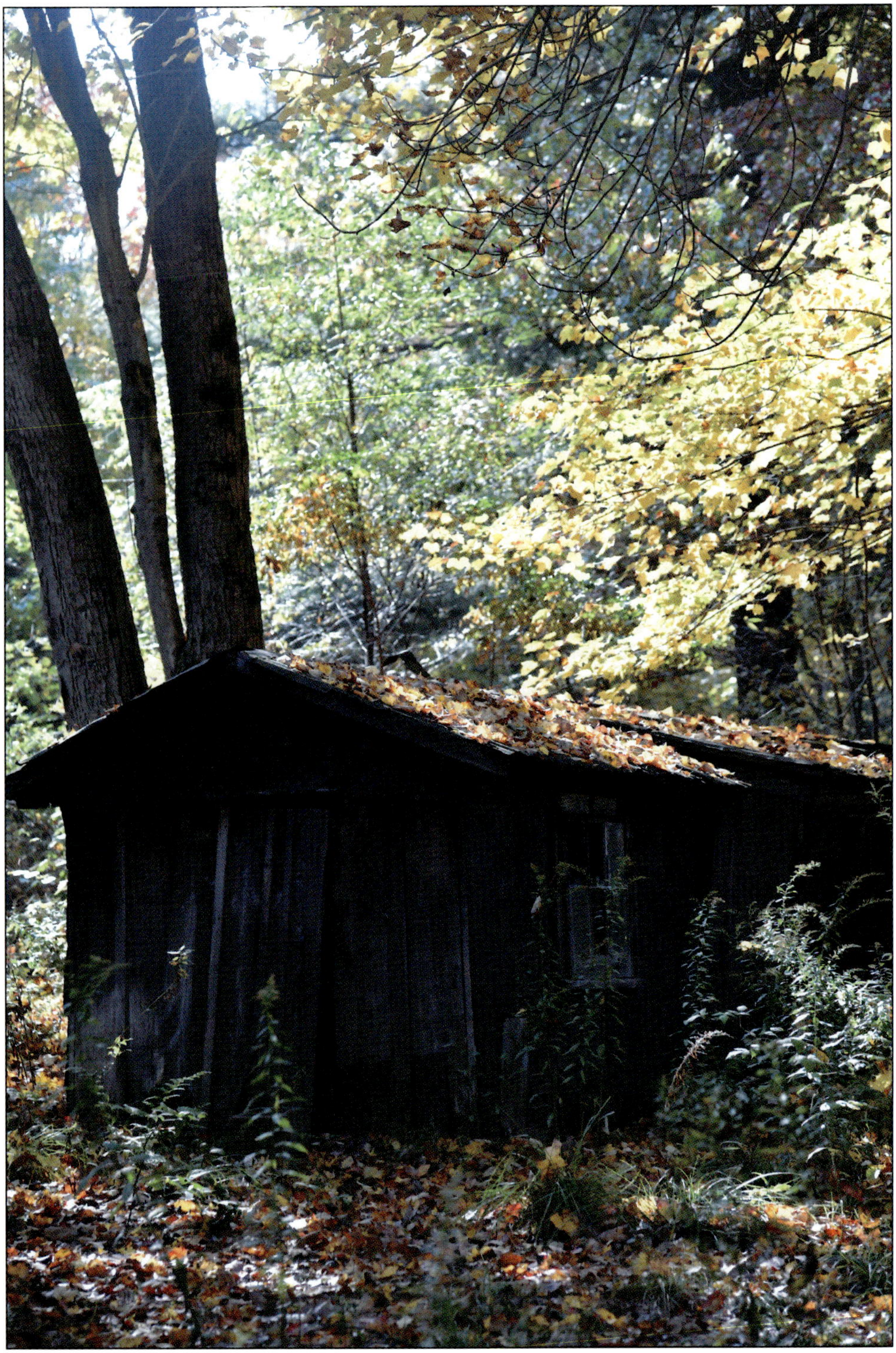

The living conditions in coal patches were substandard and bleak. There was coal heat but no running water, hence the necessity for outhouses. Their homes were as cramped as their working conditions.

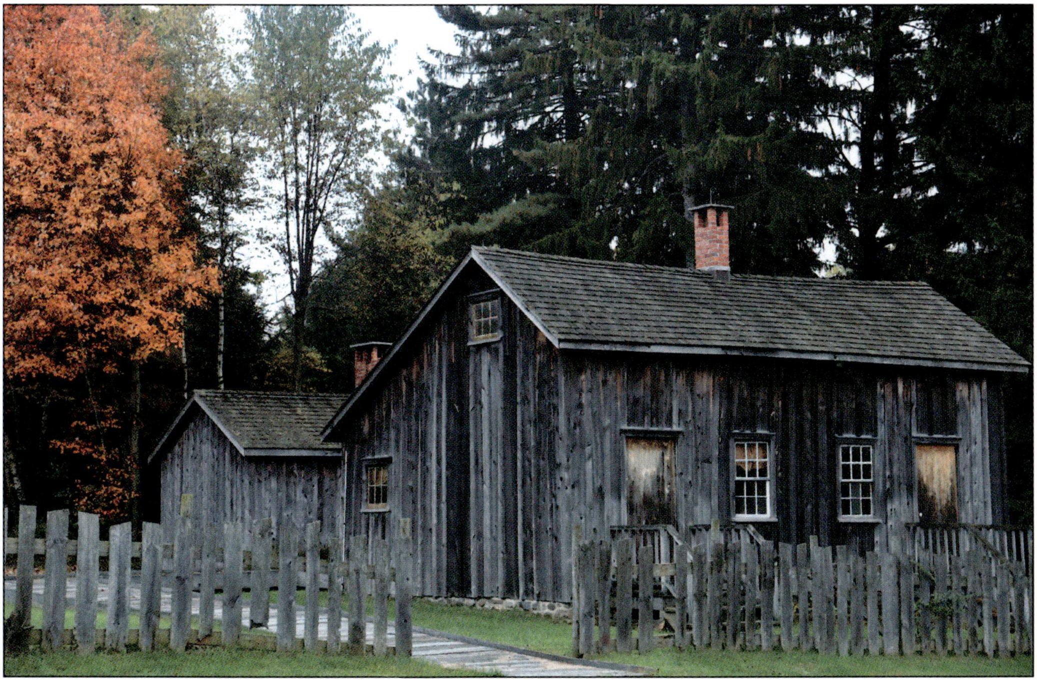

The homes pictured here housed slate pickers, jobs that were usually held by young boys or disabled men. Their jobs were the lowest on the pay scale, and as a result, they had little money for decent housing. As the smallest dwellings in the village, there are only three rooms per unit.

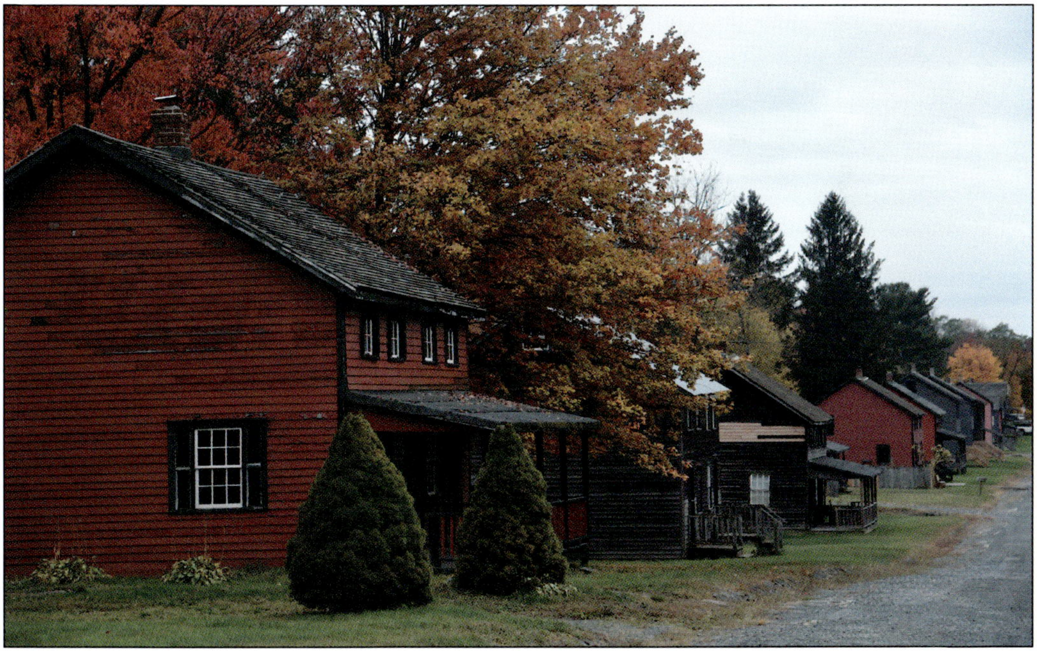

These homes are a bit larger because they housed laborers who were higher on the pay scale. Even though the homes are bigger, with four rooms in each unit, they often housed up to 15 people. A family kitchen and bedroom were on the first floor while the second floor had two bedrooms that housed boarders.

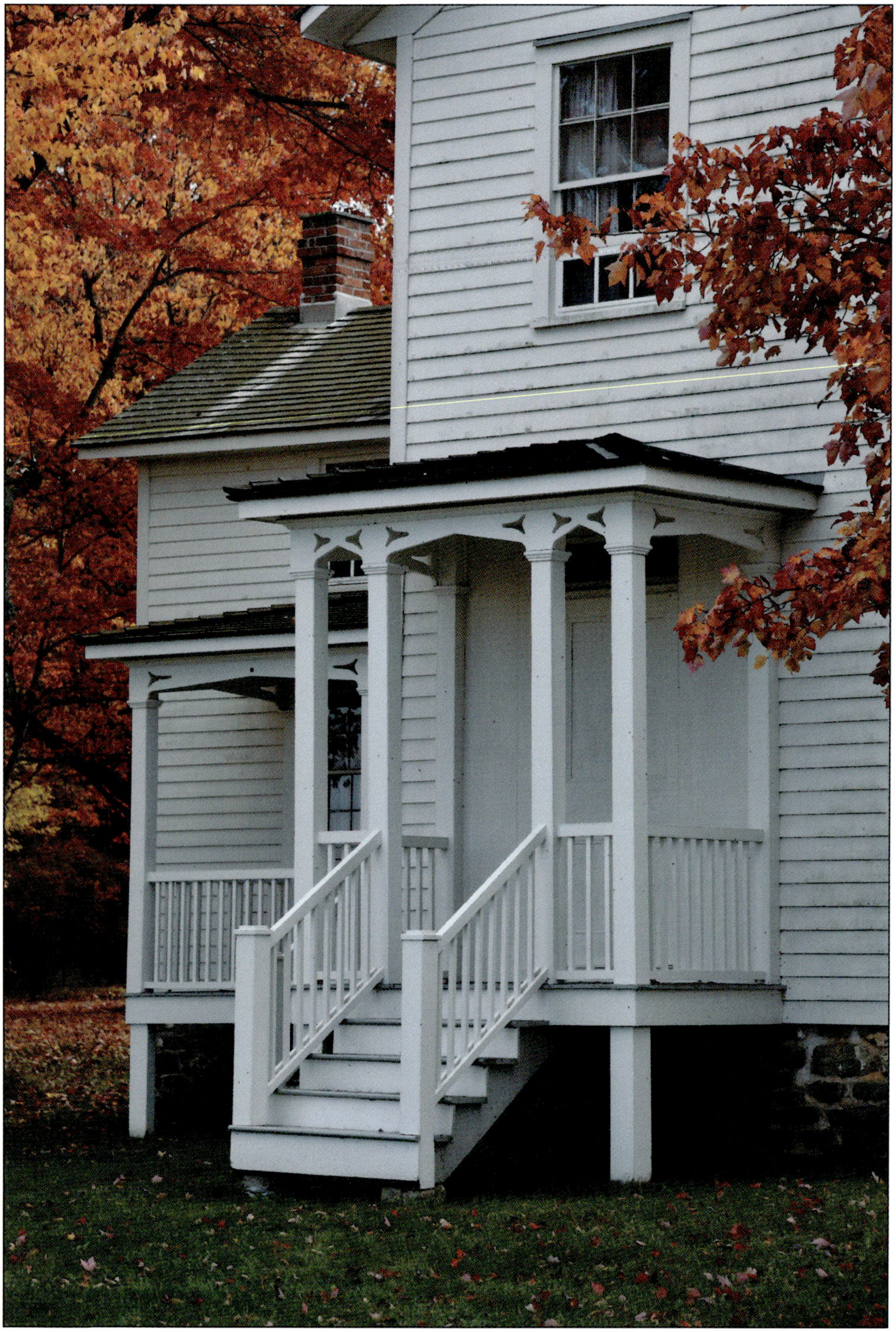

Homes like this one demonstrate the disparate living conditions in coal patches. Beautiful, roomy structures like this one would have been occupied by an employee in management. In addition to better living conditions, managers' children were afforded the opportunity of education while poorer employees' children were not.

Since the company store could be financially taxing, gardens were cultivated. Any excess food was sold for much-needed extra income or was traded. The majority of the garden work was done by women and girls.

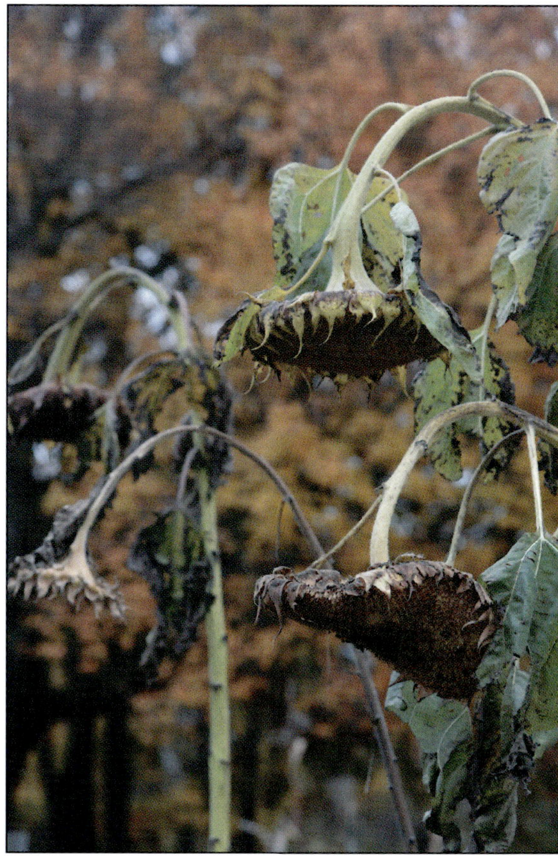

Young girls did not work in the coal mines because they had so much work to do at home. In addition to raising their siblings, they cooked, cleaned, and did laundry. In some cases, the girls assisted their mothers in doing other people's laundry for extra income.

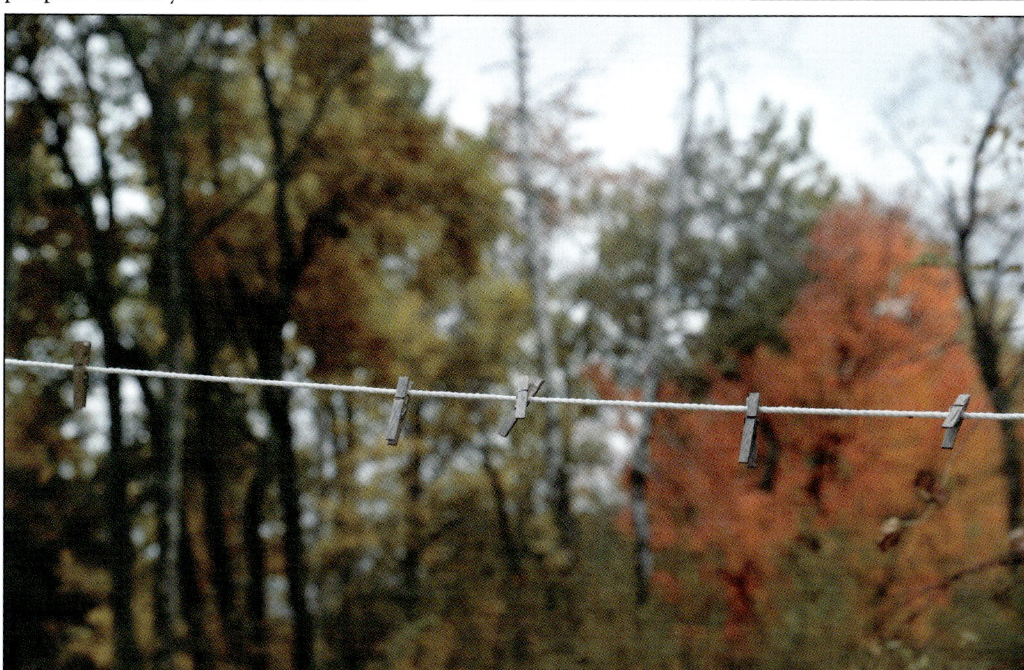

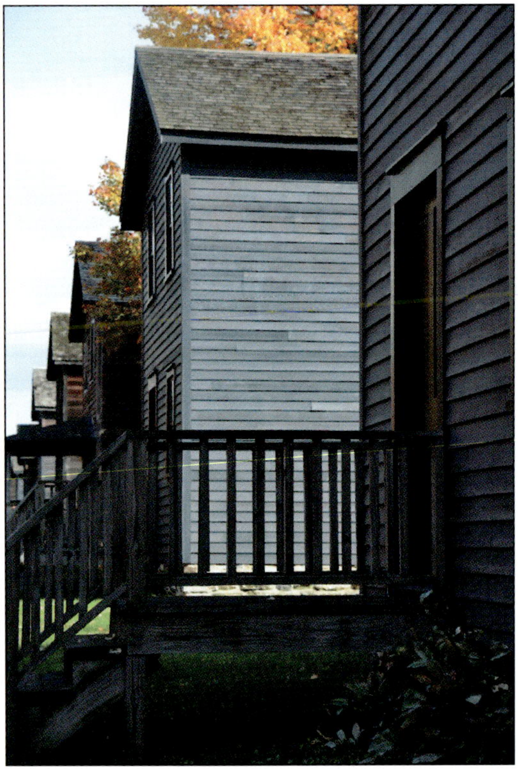

Living in a coal-patch town did not provide a great deal of security. Should a worker terminate his employment, or worse die on the job, families had to vacate their rental immediately. Often, a widow's belongings were taken from her home by coal company employees and placed in the street while she was still grieving. With no husbands to provide, those wives and their children faced a harsh future.

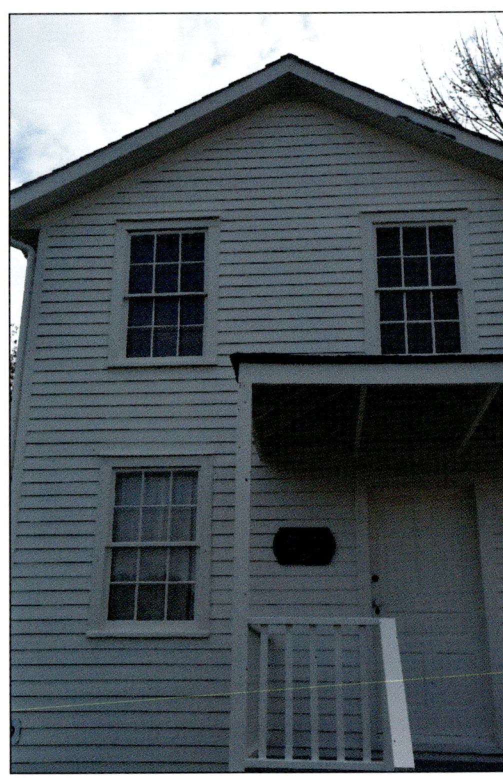

Even a simple bout with a cold could lead to serious consequences. "They had a high incidence of diphtheria, tuberculosis, pneumonia and they didn't have antibiotics," says Supey. With money in short supply, doctor visits were not a frequent occurrence, since residents had to pay the company doctor for services rendered.

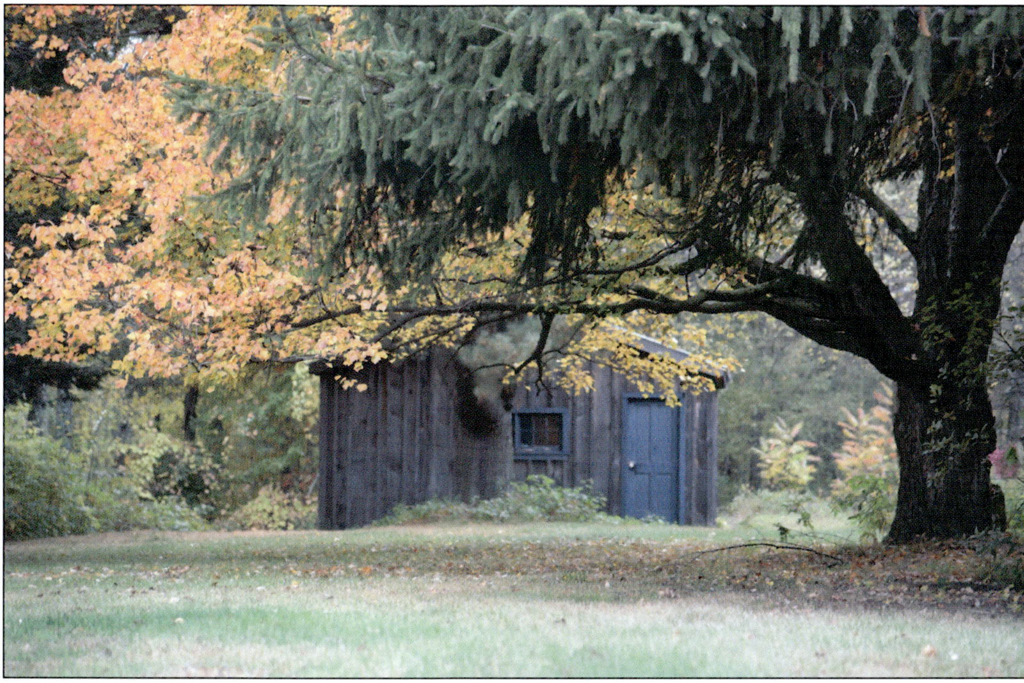

"There was no welfare. There was no free food and no low-income housing. There was none of that," says Supey. These families, who were often a continent away from their family, were often left to fend for themselves should hardship befall them.

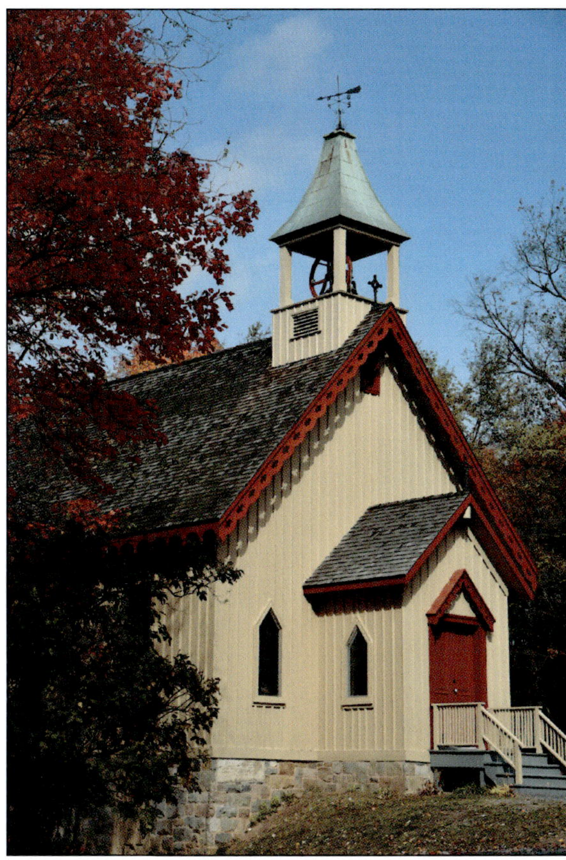

Coal-patch residents found comfort in their faith. The St. James Episcopal Church was constructed in 1859 but was later demolished. Most of its congregation was comprised of English, Welsh, and German parishioners. Pictured here, a near duplicate of the original house of worship was donated to Eckley Village in 1974, when it was relocated from White Haven.

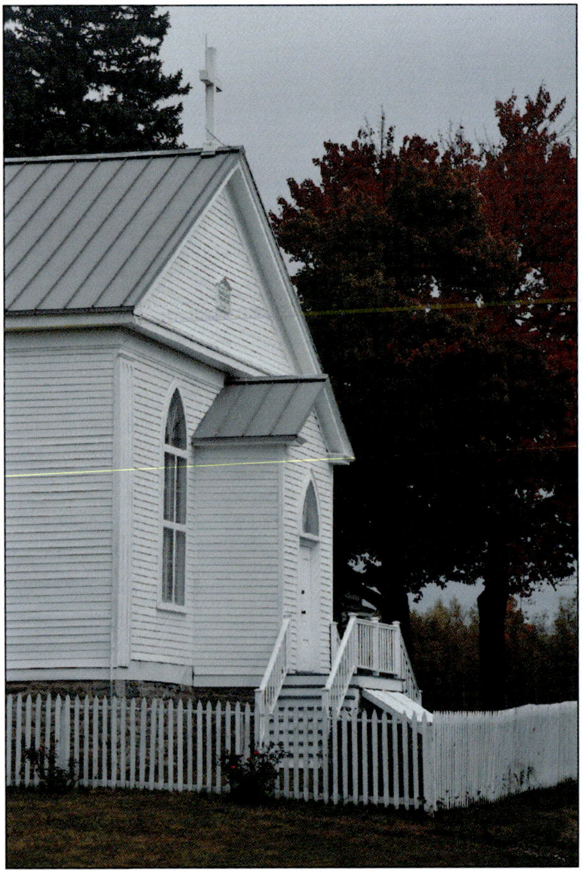

Pictured here is the original Immaculate Conception, which was a Catholic church in the poorer section of town. Built in 1861, it primarily served Irish immigrants. The church was deconsecrated in the 1950s, but the original altar still remains.

Today, Eckley Miners' Village is more than a museum. It is still home to families. "There are some people that are descendants [of original residents], but there are some people that live there just because they want to live there," says Dr. Morin. He adds that the solitude of this wooded setting has been one factor in attracting residents.

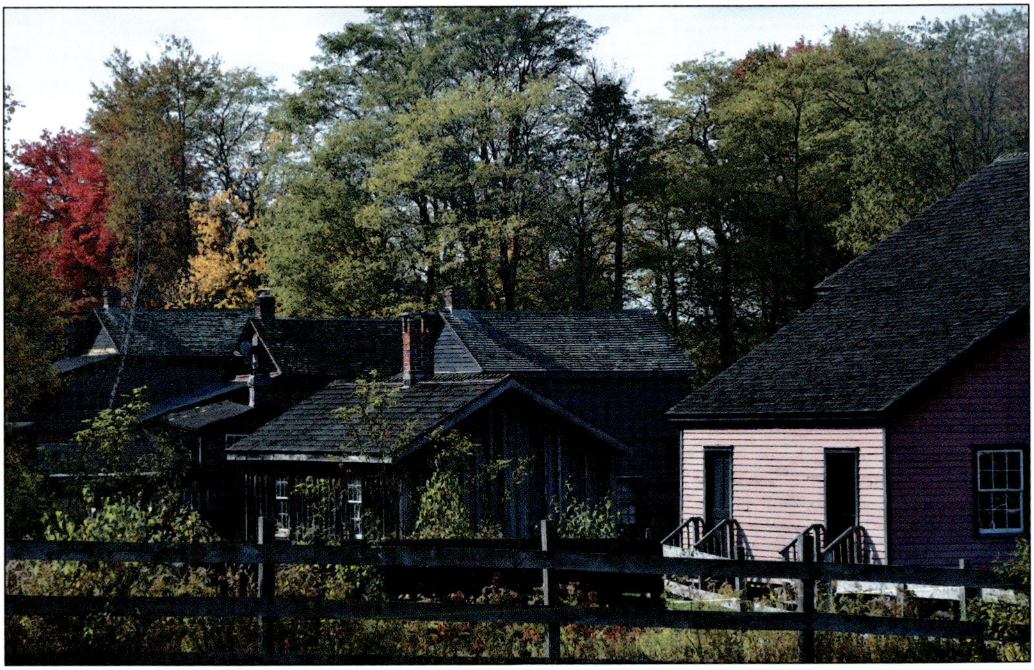

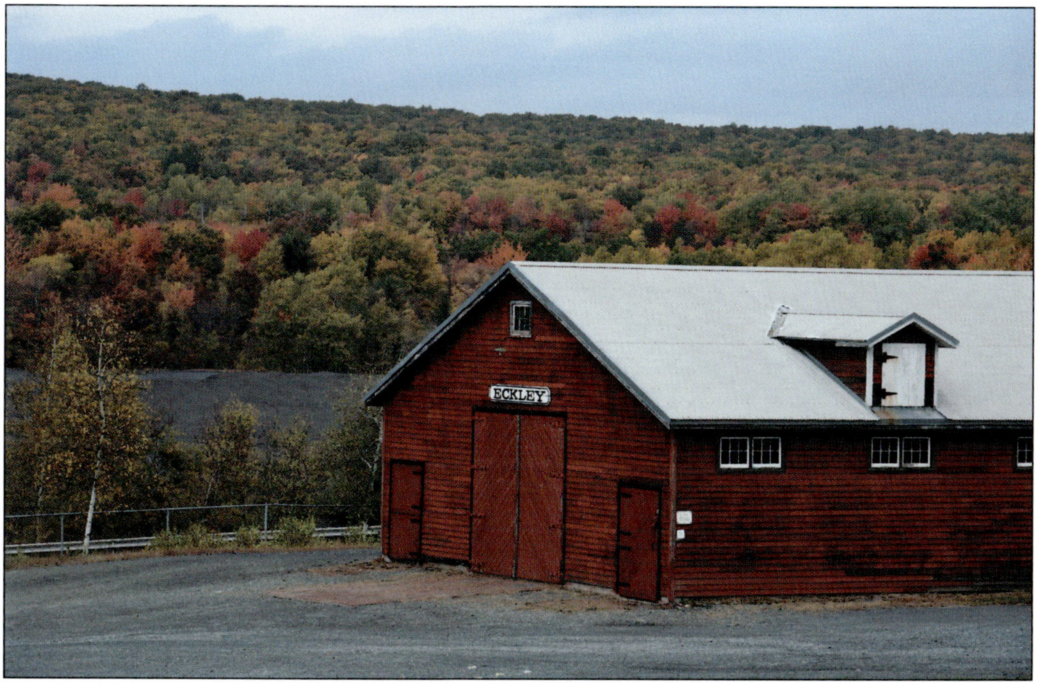

Since Eckley Village is also a functioning neighborhood, visitors should remember to be courteous and respectful of residents' privacy. For more information on Eckley Village, Eckleyminersvillage.com provides a list of upcoming events and more.

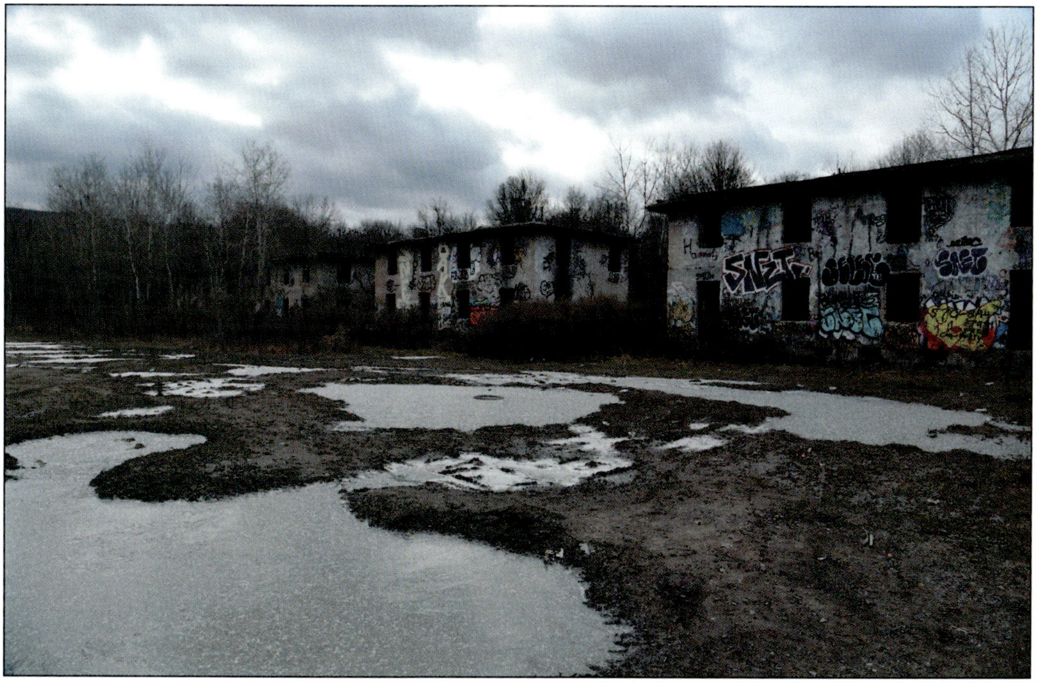

Concrete City was touted as an upscale version of the coal patch. Located in Hanover, just minutes from Wilkes-Barre, it still stands today but is an eerie reminder of a social/architectural experiment gone awry.

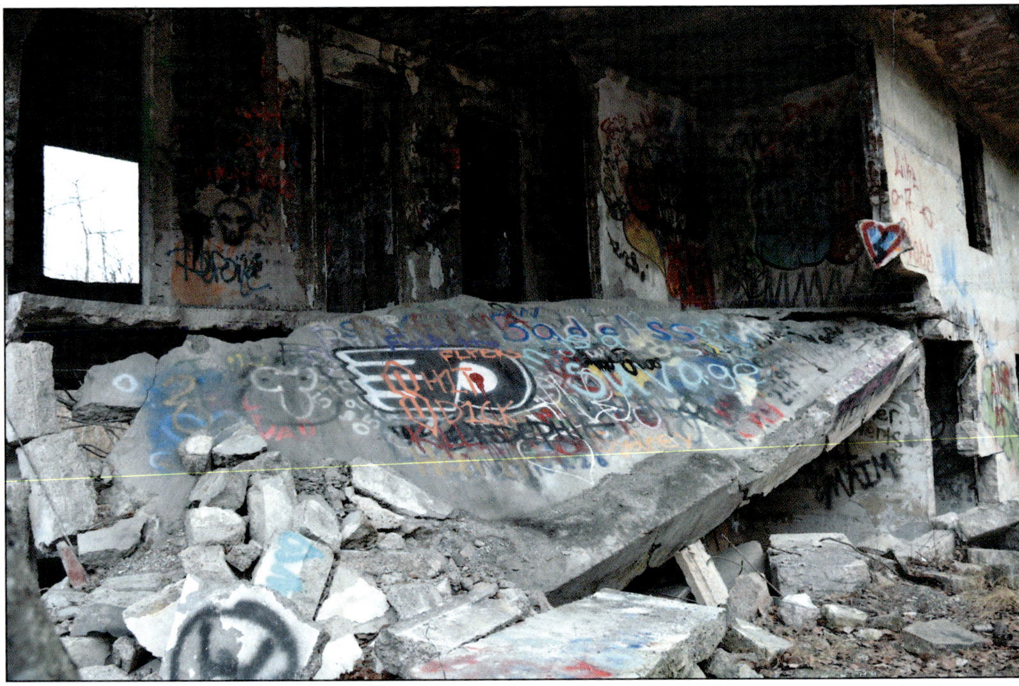

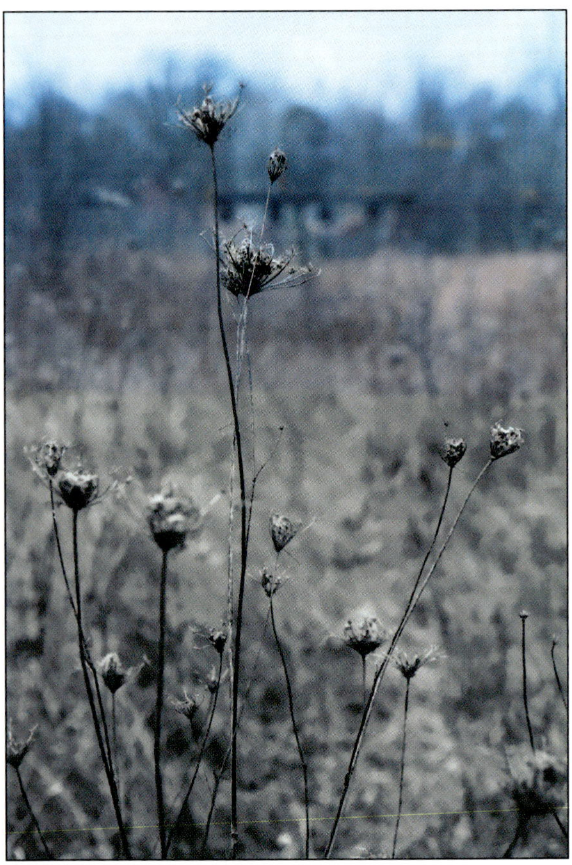

Concrete City was built in 1911, when the DL&W constructed 20 double-block structures entirely out of concrete to house railroad employees and miners. At the time, concrete was seen as a technological advance of the Industrial Revolution. DL&W executives decided it was time to start building homes from the material.

The concrete coal patch was hailed by the DL&W as the "Garden City of the Anthracite Region." Only employees with higher-echelon jobs were permitted to rent there. Those jobs included foreman, shopman, or technician. In addition, 40 men from the Truesdale mine (out of 1,700 employees) were permitted to live here.

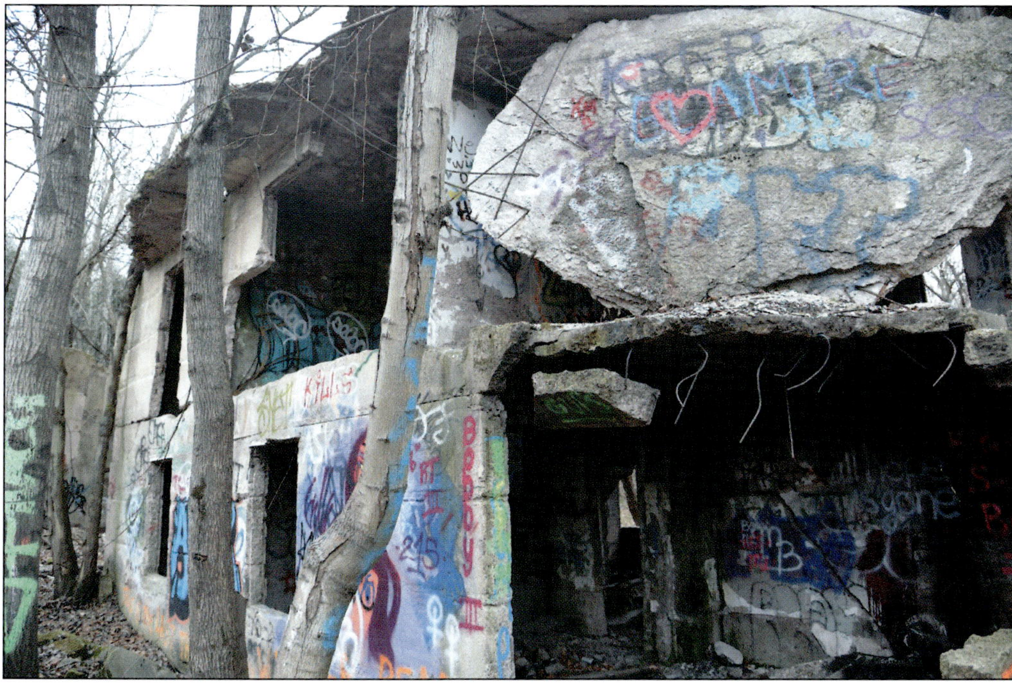

Each structure housed two families, at a cost of $8 per month. Four bedrooms occupied the top floor, with the kitchen, dining room and living room on the first floor. Concrete outhouses were located outside in the backyard.

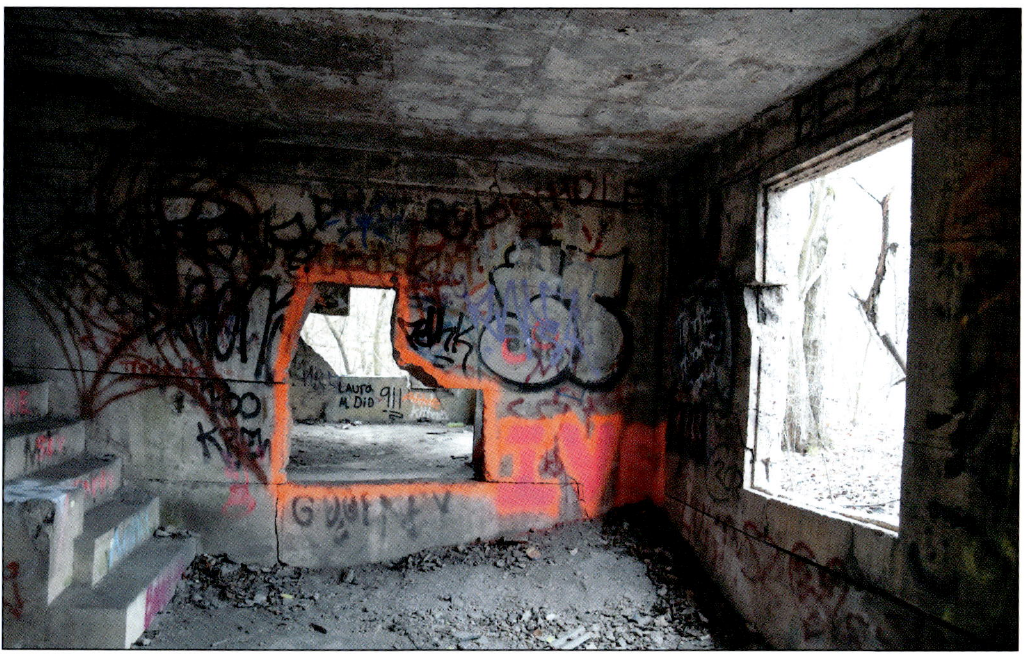

Despite high hopes for the community, Concrete City quickly began to deteriorate. A crucial construction flaw that contributed to its demise was condensation that formed on interior walls. One inhabitant reported having to thaw her father's shirts in the winter. By 1920, paint began to crack and peel. Only 13 years old, Concrete City was shuttered in 1924.

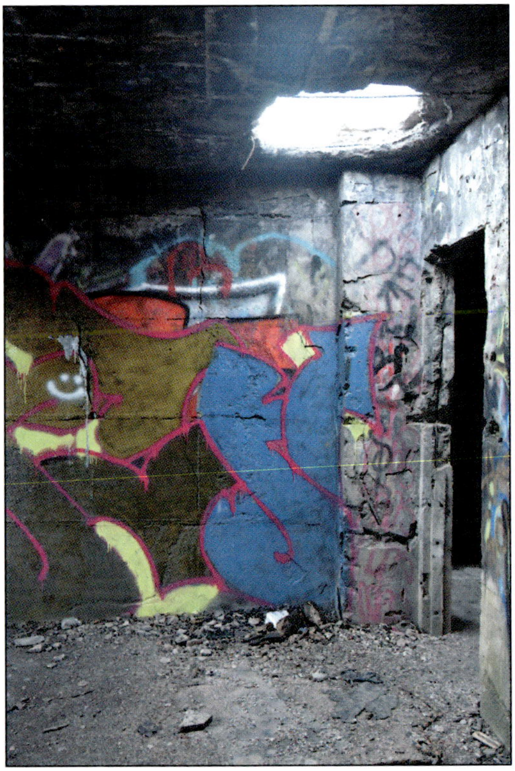

Plans to demolish the 20 buildings were squashed after the Glen Alden Coal Company purchased the land and buildings. After a hundred sticks of dynamite were placed in one home, they failed to level the structure. Today, nature has been left to its own devices to dismantle the buildings.

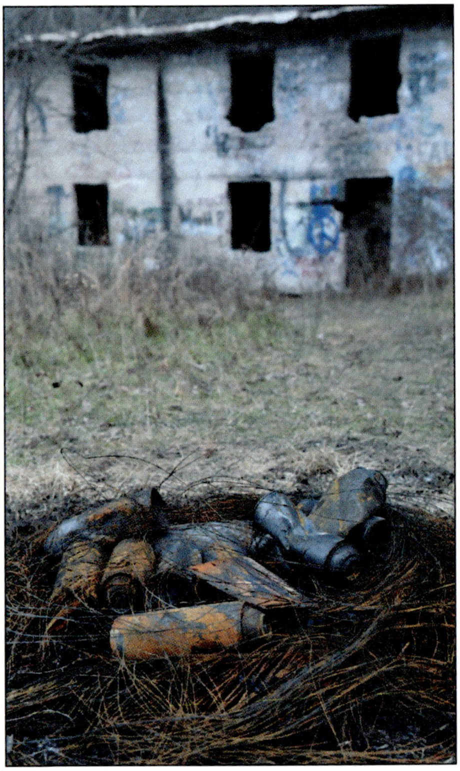

Though Concrete City is not an official tourist attraction, countless visitors continue to make the trek there. Some like to leave their marks with spray paint while others are simply content to take photographs. Those wishing to visit can follow the directions on Google maps to the gate outside of Concrete City, pass through the gate, and continue down the path.

About the Author

When Lorena Beniquez started photographing the Huber coal breaker, she had no idea it would turn into a lifelong pursuit. For over 30 years, the photographer has documented a fading history that defined not only Northeastern Pennsylvania but the United States as well. It also happens to be a part of Beniquez's family history. Her great-grandfather Stefano Pantano was a coal miner who later succumbed to black lung disease.

The Huber Breaker in Ashley was Beniquez's first anthracite subject. In 2014, the breaker was demolished just months after Beniquez last photographed it. With no muse left to photograph, she turned her camera to other pieces of anthracite history. Beniquez reached out to the owner of the St. Nicholas Breaker, the Reading Anthracite Corporation, which granted her unfettered access to document the structure.

As a writer, Beniquez has been a contributor to *Writer's Market 2016* and to the 2015 and 2014 editions of *Guide to Self-Publishing,* all published by F+W Media.

An exhibition of images from *Lost Coal Country of Northeastern Pennsylvania* has traveled throughout Northeastern and Central Pennsylvania. For information on future exhibitions, readers can visit *Lost Coal Country of Northeastern Pennsylvania* on Facebook.